D1450862

"I have been working and publishing on the topic of sustainability since the 1960s and it is always a pleasant surprise to learn something new. Orsato's book is such a surprise. It not only provides a well-grounded analysis of the rationale for more sustainable business models, but also provides fascinating stories about people and firms that have made the attempt, some successfully, some not."

—**Robert Ayres**, *Emeritus Professor, INSEAD*

"It is refreshing to find a book with an academic foundation that is easy to read and provides practical management insight. This is particularly significant when dealing with sustainability; a topic that frequently generates more heat than light. Dr Orsato, in *Sustainability Strategies*, recognises the competitive marketplace, and applies rational business decision models to meet the challenge of translating sustainability into successful business strategy. Using his detailed studies of businesses from around the globe he explores the successes and failures of firms embarking along green paths. Insightful questions are posed allowing managers to get to the heart of environmental strategy and competitive advantage, avoiding the pitfalls of seduction by feel-good factors."

—**Richard Christie**, *General Manager, Strategic Development, Ravensdown Fertiliser Co-op, New Zealand*

"In the long term, all businesses must live in harmony with their environments, but how to do so has long been the subject of anecdotal feel-good stories rather than real analysis. This book provides useful models, strategies and examples for charting a sustainable course into the future."

—**Frans Cornelis**, *Managing Director, Group Marketing & Communications, Randstad Holding, Holland*

"Does it pay to invest in IT? Yes, but only if one combines the investment with upgrading people and organizational capabilities and if one links the investment to innovation. Does it pay to be green? Yes, but again, only if this investment is linked to upgrading capabilities and to innovative business models. That is what this long overdue book shows through a set of convincing case examples put in the framework of value innovation. Being green pays and this book tells the reader how to go about making green investments profitable."

—**Luk Van Wassenhove**, *Henry Ford Professor of Manufacturing, and Academic Director of the INSEAD Social Innovation Centre*

"We now know that we cannot interpret history without a better understanding of ecology. We truly do impact on a resource-constrained world and face major issues. This very worthwhile book is for managers and academics, business leaders and specialists. It links key environmental issues and choices to hard-nosed matters of strategy and competitiveness. It breaks new ground with excellent, insightful examples from different industries and a very realistic approach."

—**Anthony Simon**, *former President of Marketing, Unilever; member of the Board of IIIEE, Lund University and of INSEAD's Alumni*

"These days we all profess to support sustainability, and many organizations want 'to do the right thing'. But how can managers make decisions to guide them through the maze of options for eco-investments, stay focused and become greener? They can read this timely book!"

—**Stewart Clegg**, *Professor of Management and Research Director of the Centre for Management and Organization Studies, University of Technology, Sydney, Australia*

"Orsato's global background provides a sweeping view of the many types of successful sustainability strategies firms have employed around the world. Each chapter focuses on one of Orsato's sustainability strategies, and each strategy comes with a useful checklist of questions managers need to address before pursuing it."

—**Charles J. Corbett**, *Professor of Operations Management and Environmental Management, UCLA Anderson School of Management, Los Angeles, US*

"As we confront climate change and move toward the low-carbon economy, this book will be an invaluable guide for executives who seek to align environmental and social investments with the strategy of the enterprise. Orsato has taken a highly strategic approach and created a systematic, theoretically based but practical guide as to where to make investments for sustainability strategy which will both benefit the planet and enhance competitiveness. The argument is clearly laid out, logically developed and presented in clear, accessible English. I recommend it highly for reading by both executives and students in MBA strategy courses."

—**Dexter Dunphy**, *Visiting Professor, University of Technology, Sydney, Australia*

# SUSTAINABILITY STRATEGIES

**INSEAD Business Press Series**

J. Frank Brown
THE GLOBAL BUSINESS LEADER
Practical Advice for Success in a Transcultural Marketplace

Lourdes Casanova
GLOBAL LATINAS
Latin America's Emerging Multinationals

David Fubini, Colin Price & Maurizio Zollo;
MERGERS
Leadership, Performance and Corporate Health

Manfred Kets de Vries, Konstantin Korotov & Elizabeth
Florent-Treacy
COACH AND COUCH
The Psychology of Making Better Leaders

Renato J. Orsato
SUSTAINABILITY STRATEGIES
When Does It Pay to Be Green?

James Teboul
SERVICE IS FRONT STAGE
Positioning Services for Value Advantage

Jean-Claude Thoenig & Charles Waldman
THE MARKING ENTERPRISE
Business Success and Societal Embedding

Rolando Tomasini & Luk Van Wassenhove
HUMANITARIAN LOGISTICS

# SUSTAINABILITY STRATEGIES

When Does It Pay to Be Green?

Renato J. Orsato

First published 2009 by
PALGRAVE MACMILLAN

Palgrave Macmillan in the UK is an imprint of Macmillan Publishers Limited, registered in England, company number 785998, of Houndmills, Basingstoke, Hampshire RG21 6XS.

Palgrave Macmillan in the US is a division of St Martin's Press LLC, 175 Fifth Avenue, New York, NY 10010.

Palgrave Macmillan is the global academic imprint of the above companies and has companies and representatives throughout the world.

Palgrave® and Macmillan® are registered trademarks in the United States, the United Kingdom, Europe and other countries.

ISBN-13: 978–0–230–21298–5
ISBN-10: 0–230–21298–0

This book is printed on paper suitable for recycling and made from fully managed and sustained forest sources. Logging, pulping and manufacturing processes are expected to conform to the environmental regulations of the country of origin.

A catalogue record for this book is available from the British Library.

A catalog record for this book is available from the Library of Congress.

10   9   8   7   6   5   4   3   2
18   17   16   15   14   13   12   11   10   09

Printed and bound in Great Britain by
CPI Antony Rowe, Chippenham and Eastbourne

*This book is dedicated to the memory of my beloved mother, Olinda Romani Orsato (Nov. 10, 1939–Dec. 3, 2004), an unconditional supporter of all my intellectual journeys, whose ability to love and accommodate conflict inspired my optimism – that a sustainable world is indeed possible.*

# CONTENTS

CONTENTS

# LIST OF FIGURES

# LIST OF ACRONYMS

| | |
|---|---|
| AESC | Automotive Energy Supply Corporation |
| AFPA | American Forest and Paper Association |
| APV | All-Purpose Vehicles |
| B2B | Business to Business |
| B2C | Business to Consumers |
| BART | San Francisco Bay Area Rapid Transit District |
| BELC | Business Environmental Leadership Council |
| BMU | German Federal Ministry of the Environment |
| BOS | Blue Ocean Strategy |
| $CaCO_3$ | Calcium Carbonate |
| CCPA | Canadian Chemical Producers Association |
| CDM | Clean Development Mechanism |
| CERES | Coalition for Environmental Responsible Economics |
| CES | Competitive Environmental Strategies |
| CF | Carbon Footprint |
| $CH_4$ | Methane |
| CMA | American Chemical Manufacturers Association |
| CMS | Chemical Management Services |
| $CO_2$ | Carbon Dioxide |
| CSOs | Car-Sharing Organizations |
| CSP | Chemical Strategies Partnership |
| CSR | Corporate Social Responsibility |
| DDCE | DuPont Danisco Cellulosic Ethanol |
| DJSI | Dow Jones Sustainability Index |
| EC | European Commission |
| EIPs | Eco-industrial Parks |
| ELVs | End-of Life Vehicles |
| EMS | Environmental Management Systems |
| EPR | Extended Producer Responsibility |
| ETS | Emission Trading Scheme |
| EU | European Union |

| | |
|---|---|
| EVs | Electric Vehicles |
| FDA | Federal Drug Administration |
| FM | Food Miles |
| FSC | Forest Stewardship Council |
| GE | General Electric |
| GHG | Greenhouse Gasses |
| GM | General Motors |
| GMOs | Genetically Modified Organisms |
| GPS | Global Positioning System |
| GRAS | Generally Recognized as Safe |
| GRI | Global Reporting Initiative |
| HCCP | Hazardous Critical Control Points |
| HDPE | High Density Polyethylene |
| HFCs | Hydrofluorcarbons |
| HV | Hybrid Vehicles |
| ICEs | Internal Combustion Engines |
| IE | Industrial Ecology |
| IPCC | Intergovernmental Panel on Climate Change |
| IPONZ | Intellectual Property Office of New Zealand |
| IPR | Individual Producer Responsibility |
| IS | Industrial Symbiosis |
| ISO | International Organisation for Standardization |
| JI | Joint Implementation |
| KRAV | Swedish eco-label for organic food |
| LCA | Life-Cycle Assessment |
| LT | Lean Thinking |
| MCC | Micro Compact Car |
| MNC | Multinational Corporations |
| MSS | Mobility Service Systems |
| $N_2O$ | Nitrous Oxide |
| PCA | Partnership for Climate Action |
| PFCs | Perfluorcarbons |
| PIVCO | Personal Independent Vehicle Company |
| PP | Polypropylene |
| PSS | Product-Service Systems |
| PS | Positioning School |
| PVC | Polyvinylcloride |
| R&D | Research and Development |
| RAN | Rainforest Action Network |
| RBV | Resource Based View |
| SFI | Sustainable Forest Initiative |

| | |
|---|---|
| SPP | Southern Pacific Petroleum |
| SVI | Sustainable Value Innovation |
| TGC | Tradable Green Certificates |
| TIMM | Terrestrial Individual Motorized Mobility |
| TRM | Total Responsibility Management |
| TQM | Total Quality Management |
| TWC | Tradable White Certificates |
| UK | United Kingdom |
| UNCED | United Nations Conference on Environment and Development |
| UNEP | United Nations Environmental Program |
| UNFCCC | United Nations Framework Convention on Climate Change |
| US | United States |
| USCAP | United States Climate Action Partnership |
| USPCSD | US President's Council for Sustainable Development |
| VDA | German Automobile Industry Association |
| VEI | Voluntary Environmental Initiatives |
| VOC | Volatile Organic Compounds |
| WASD | Weighted Average Source Distance |
| WBCSD | World Business Council for Sustainable Development |
| WEEE | Waste Electrical and Electronic Equipment |
| WWF | World Wildlife Fund |
| ZERI | Zero Emissions Research Initiative |
| ZEV | Zero Emission Vehicles |

# PREFACE

Environmentalism is subject to cycles of perception and interest. During the 1970s and 1980s, alarmist peaks of scarcity, pollution and biodiversity loss triggered only sporadic attention to ecological issues by the media, politicians and business. Environmentalism struggled to survive. The United Nations Conference on Environment and Development (UNCED) held in Rio de Janeiro, Brazil in 1992 started to change both the political agenda and the positioning of some large corporations. Environmental issues started to be treated in a more serious and systematic manner, and the remaining part of the decade was marked by intense activity in both practice and theorization. A large number of companies implemented tools, standards and reporting initiatives for proactive environmental management, often with help of consultants and in partnership with Non-governmental Organizations (NGOs). In the academic front, the decade was dominated by a heated debate about whether it pays to be green, which, not surprisingly, ended with no winners. As the end of the 20th century approached, empirical evidences of environmental protection and restoration were scarce, while the debate about the value of eco-investments tired academic circles, loosing momentum. The result was another downturn for environmentalism.

In the early years of the new millennium, the focus shifted from environmental to social issues. An increasing number of books, academic papers and popular articles emphasized the social dimension of business responsibility and endorsed the need and advantages of good corporate citizenship. Prescriptions were justified more with carrots than sticks. According to the legions of proponents of Corporate Social Responsibility (CSR), positive rewards would not only outperform penalties but leadership in social causes would generate competitive advantages. In other words, the *pays to be green* debate had migrated to the realms of social responsibility. The issue moved to *does it pay to be nice*? This time, however, the discourse was more appealing. Besides the proclaimed business advantages, who would argue against societal demand for corporate accountability? From the point of view of citizenship, who would deny the need for corporations to be sensitive to human rights? In free and democratic societies, advocating against political correctness would simply be an oxymoron. The result was an overwhelming early success of the CSR discourse.

Corporate scandals in the United States (US) and Europe in the early 2000s, further sensitized public opinion and reinforced the business case for more ethics and governance in business. As CSR took the stage, there was a massive migration of scholars and consultants from the old and boring field of environmental management, to social issues in business. They proposed that environmental management should be under the umbrella of CSR, and be treated in the same way. This also caused a shift in focus of the corporate world. Instead of showing how they managed waste, for instance, corporations were keener in presenting their efforts to tackle social problems. Suddenly, the environment could speak, making CSR reports sexier than the former environmental accounts. Corporate representatives could collect testimonials from stakeholders, something almost impossible in the area of environmental protection. After all, air, water, soil and biodiversity do not speak and hardly attract any positive media attention, as people do. Because nature has no face, it is easily forgotten.

After much media attention, empirical evidence that CSR pays remained scarce,[1] causing some to start questioning the normative discourse. The ratification of the Kyoto protocol in the end of 2004 contributed to further change the political climate. More literal, however, were the escalation of climate disruptions influenced by global warming, such as the Hurricane Katrina in New Orleans in 2005. The personal crusade of Mr Al Gore, depicted in the *Inconvenient Truth* documentary, and the 2007 Nobel Peace Prize (announced in December 2006), legitimized climate change as the most pressing issue for humankind in the new millennium. By then, the anti-Kyoto standing of the George W. Bush presidency in the US started to wear out, and public demand for actions brought environmental issues to the center of political agenda, now consolidated in the term *sustainability*.[2] Candidates running for public positions, corporate executives and deans of business schools, suddenly started claiming their environmental credentials. Whatever caused the tipping point, the pendulum brought back the urgent need for societies to reduce the anthropogenic impacts on the natural environment, pointed out in the 1970s.

As the pendulum swung to the sustainability side, it brought two curious outcomes. First, many believed that climate issues could be treated as another type of CSR issue, which so far was mostly peripheral to business. While social and environmental issues are two sides of the same coin – the coin itself representing economic development – quite often social and environmental issues require different treatment. As Porter and Reinhardt[3] emphasized: "Companies that persist in treating climate change solely as a corporate social responsibility issue, rather than a business problem, will risk the greatest consequences". Confronted with broad spectra of CSR and sustainability-driven initiatives, corporations still have to base their investments on sound rationales for competitiveness. Most of them, however, still do not know what to do and why. Although nowadays there is more awareness, by the end of the 2000s, a great number of executives are still uncertain what sustainability means for their businesses and, more important, what they can do about it.

Second, from the pessimistic perspective that greening *never* pays that dominated both the mindset of business and the specialized literature until quite recently, the pendulum moved to the other end: to the view that greening *always* pays. Unfortunately, from both the business and the environmentalist perspectives, this book will show that the scope for win–win scenarios is narrower than many wish them to be. Out of the vast array of actions taken by firms, only a few will be profitable, generate competitive advantage or create new market spaces. This is not, however, all bad news. After all, sustainability is similar to any other issues in business: the profitability of investments depends on the internal and external contexts of the organization. Therefore, to restore the balance between pro and against greening arguments, we need to identify when and why environmental investments pay.

The argument for conditional returns does not imply companies should do little to protect natural ecosystems. On the contrary, firms should undoubtedly deploy increasingly ambitious social goals and do what eco-activists and the general public expect: the reduction of the overall environmental impact of businesses, so that sustainable societies can eventually emerge. However, if strategy is about "doing better by being different", as Joan Magretta[4] elegantly put, then sustainability strategies require more than doing well. Managers need to identify the appropriate types of eco-investments to focus their efforts in the pursuit of competitive advantage or the creation of new market spaces. They will need to identify *when it pays to be green*. By revisiting this recognizably ample question, as well as others left unanswered by the specialized literature, this book redirects the attention to the importance of finding business rationales for sound *corporate environmentalism* – a term used in this book to represent the broad range of practices that have the potential to reduce the impact organizations cause to the natural environment.

The analytical reasoning and frameworks presented in the book can help managers to prioritize eco-investments according to the fundamentals of strategic management. Executives who are already convinced that they should do something about sustainability will find important insights in the book. *Sustainability Strategies* will help them to apply business principles to environmental and social issues, so as to increase the overall competitiveness of their firms. Through the systematic use of analytical frameworks, the book connects the variables implicated in the formation and evaluation of sustainability strategies. The frameworks also help managers to define and prioritize areas of organizational action and optimize the economic return on eco-investments. By doing so, managers can more easily align the sustainability strategy with the generic business strategy.

The book is based on analytical reasoning and studies of individual organizations and groups of organizations. Most case studies result from extensive research with a global range. Nonetheless, in order to avoid overloading the reader with scattered or unrelated examples, the cases used throughout the book are limited to those that help grounding theory. The presentation of an exhaustive number of successful cases, as many books on the topic do, is crucial to make the business case for corporate environmentalism. However, to a large extent, the business case has long been presented. Reasonably well-informed executives are already convinced that sustainability is indeed a core business issue. What is necessary now is to connect apparently disparate cases via grounded theoretical frameworks. Finally, the book not only builds on success stories but it also presents several examples of eco-investments that simply failed. Such experiences are equally important for managers to learn from when it does *not* pay to be green.

The book is divided into three parts. The Chapters 1 and 2 forming Part I (Fundamentals) explore the two fundamental research questions addressed by the book. Chapter 1 shows that answering *when it pays to be green* requires an understanding of the scope of corporate environmentalism, and the elements involved in the so called win–win scenarios of simultaneous economic and environmental gains. Overall, the conditions favoring or hindering returns on eco-investments have a plethora of methodological difficulties, addressed in the final sections of the chapter. Chapter 2 reveals the main differences between operational effectiveness and strategy in the realms of corporate environmentalism, by analyzing what sustainability strategies are. The chapter also presents the analytical frameworks that will serve as the backbones of the remaining chapters of the book. Four sustainability strategies used by firms to compete in existing industries are complemented by a fifth strategy used by companies to create new market spaces while addressing the demands for environmental and social responsibility.

Part II (Competitive Environmentalism) includes Chapters 3 to 6 wherein it discusses the four Competitive Environmental Strategies (CES). Chapters 3 and 4 address the strategies focusing on organizational processes, while Chapters 4 and 5 present the strategies in which the focus is on the products and services sold by the company. By exploring three areas of potential gains – lean thinking, industrial ecosystems and carbon credits – Chapter 3 presents the rationales for firms to adopt Eco-efficiency as a specific sustainability strategy. Chapter 4 delves into the intangible value of eco-investments, and identifies when positive reputation is prone to emerge for companies that join voluntary environmental initiatives or, as some call, Green Clubs. Chapter 5 addresses eco-labeling schemes as means of differentiating products and services. Eco-branding strategies, however, require more than eco-labeling, as the analysis of cases from Sweden and Australia demonstrate in the final part of the chapter. Finally, Chapter 6 analyzes the ways in which companies can be the leaders in both low costs and low environmental impact of their products and services.

Part III (Beyond Competition) of the book includes Chapters 7 and 8. Chapter 7 presents the fifth and most challenging sustainability strategy. The logic embedded in concept of Sustainable Value Innovation (SVI) is uncovered via the analysis of several cases within the area of individual motorization. By delving deeply into one single topic, which is more directly related to the global automobile industry, the chapter explores the elements and subtleties involved in deploying SVI strategies. Finally, Chapter 8 concludes the book by revisiting the main benefits brought by the eco-investments and corresponding sustainability strategies analyzed in the preceding chapters.

Although the book presents a very comprehensive set of initiatives in the area of corporate environmentalism, it has some self-imposed limitations. First, *Sustainability Strategies* focuses on the choices managers have when deciding about eco-investments and their alignment with the general strategy of the business; it does not intend to be a handbook of environmental management or CSR. There are a wide range of useful practice-oriented books[5] and handbooks available to managers to identify environmental aspects and impacts, as well as practical approaches and tools to address them.[6] These materials also tend to cover the upstream (supply chain management) and downstream (product stewardship) activities, which are treated here as areas of strategic influence, rather than direct components of sustainability strategies. Second, *Sustainability Strategies* focuses on the elements involved in the development of strategy and its evaluation (also treated as strategy formation or definition). Since the implementation and control of sustainability strategies use similar logic of generic business strategies, there is not shortage of material to choose from.[7] The endnotes of each chapter identify these supporting materials, as well as explanations about the various topics treated in the book, so readers interested in investigating them further find useful help from the notes provided in each chapter.

Of course, there are limitations that I would rather avoid. My research training, interests, values and heuristics might have influenced the perceptions constituting the reality expressed in this work. Although I tried to discipline and reduce interference from personal idiosyncrasies in the development of the study, the results represent, ultimately, my personal interpretation of reality, with all the cognitive faults that others might detect therein. For those faults, I take full responsibility.

# ACKNOWLEDGEMENTS

The journey culminating in this book would not have been possible without the Marie Curie Fellowship award, which allowed me to immerse myself in broad international research during 2004–2007. I am truly grateful for the grant from European Commission (EC, Research DG Human Resources and Mobility) and the support of the commissioners Roberto Santorello and Joelle Lardot. The Fellowship was crucial to undertake a truly global project, which also made possible to work with people from four continents. Back in 2003, Landis Gabel and Kai Hockerts provided the initial support for the Fellowship application at INSEAD in Fontainebleau, France. Michelle Duhamel, Alison James and Anne Fournier were more than helpful with their administrative assistance during the duration of the Fellowship.

I am grateful for the support of the INSEAD Social innovation Centre in the development of this book. The centre is a platform for cross-disciplinary research in Social innovation and aims at introducing new business models and market-based mechanisms that deliver sustainable economic, environmental and social prosperity. (For more information about the centre's work please visit: www.insead.edu/isic).

Frank Brown, the INSEAD Dean, Stephen Chick from the R&D Committee, as well as Luk Van Wassenhove and Megan Pillsbury, the Directors of the INSEAD Social Innovation Centre, provided the crucial support for the completion of the work. In fact, Luk's support goes back to the first of the three summer holidays I spent at the INSEAD campus working on the book. His early feedback on the first chapters and case studies influenced the development of the project decisively. At the Centre, I also had the privilege to count with the wisdom of Professors Robert Ayres and Paul Kleindorfer, whose experience in environmental and risk matters transformed every conversation into a private lesson. From Bob Ayres and Benjamin (Ben) Warr, I finally learnt something that made sense about the relationship between economic growth and the energy efficiency of economies, which sooner or later, I am sure, will be become a reference for those who are serious about making sustainable development a reality.

*Sustainability Strategies* also confirm how influential the discussions with Professor W. Chan Kim about the interface between Blue Ocean Strategy and ecological sustainability had on me. Value innovation acquired a value in itself; besides a means of creating new market spaces, it inspired me to look into the possibilities of aligning business interest with the generation of public benefits. The frequent intellectual exchanges with Ben Warr has also been of great value in this respect; as we worked on the theoretical foundations of this new research endeavor, value has been literally transformed into Sustainable Value Innovation (SVI). The work of Ben and Bob Ayres is partially reflected on the bridge I attempted to make in Chapter 7 between societal, industry and organizational levels of analysis, so to anchor the (admittedly) early notions of SVI in the physicality of the economy. The teachings of David Vogel at INSEAD during the springs of 2006 and 2007 have also influenced my work. David and Craig Smith also provided me important feedback in early versions of the chapters. I also appreciate the support of the staff at INSEAD Social Innovation Centre, in particular the priceless research support of Sophie Hemne, who has been not only fantastic with the collection of secondary data but also helping me to put the pieces of the puzzle together. *Merci beaucoup* to all of you!

This more recent work at INSEAD is, in fact, the finale of a much longer journey. During the outgoing phase of my Fellowship in Australia (July 2004–2006), Anne Ross Smith (Head of the Business School), and Stewart Clegg (Director of the Centre of Innovative Collaborations, Alliances and Networks) gave me the institutional backing at the University of Technology, Sydney (UTS). The welcome (back) from Anne and Stewart, as well as other colleagues from UTS, went far beyond the office space and access to data. Cleo Lester provided me with the necessary assistance in her friendly Brazilian style, while the hospitality of Tyrone Pitsis, Ray Gordon, Rosie Stilin, Slava Konovalov, David Bubna-Litic, Jane Marceau, Antoine Hermens, John Crawford, Jochen Schweitzer, Jordan Louviere and – in and out from Mexico – Salvador Porras, also made all the difference in my stay in Sydney. The frequent interactions I had with Dexter Dunphy and Suzanne Benn over a *café late* were of great value and much beyond the average academic camaraderie. Dexter and Sue also facilitated access to research programs and companies, and quite often we found ourselves debating sustainability issues in the lobby of corporations such as Lend Lease or

Fuji Xerox. The assistance of Andrea North-Samardzic and Robert Perey in the development of the Australian case studies was also critical. Chris Ryan (Director of the Australian Centre for Science Innovation and Society, University of Melbourne) was another strong supporter of both my intellectual endeavors and field trips in Victoria and South Australia, ranging from manufacturers of furniture to wine makers in the Barossa Valley. For all your support, *thank you, mates!*

In New Zealand, Delyse Springett (Director, Programme for Business and Sustainable Development, Massey University), facilitated my research by putting me in contact with the members of the Corporate Leadership Group (CLG). The CLG seminar series, developed during 2005–2006 in Wellington and Christchurch, allowed me to gain knowledge directly from the executives of corporations such as Ravensdown Fertilizer Cooperative, Fulton Hogan, Coca Cola Amatil NZ, Watercare Services, Shell, Mobil Oil NZ, Siemens Energy Services, BP Oil NZ, and Auckland International Airport. Delyse organized several company visits, which also included touring the beautiful countryside of New Zealand to interview winemakers. The development of the *eco-n* case, described in Chapter 6, also demanded a few trips to Christchurch and surroundings. Richard Christie (General Manager, Strategic Development) from Ravensdown has been a very patient instructor, teaching me pastoral agriculture by literally taking me to the fields and factories in the Southern Island. Ron Pellow (Business Development Manager, *eco-n*) has also been a frequent collaborator in discussions and internal seminars at Ravensdown. To all my *Kiwi* friends, *Kia ora!*

The automobile industry has also exerted a strong influence in my thoughts about sustainability issues, as many parts of this book suggest. From the early days of my doctoral studies, back in the 1990s, through the very last minute corrections of the sections dealing with the auto industry in the book, I have counted with the profound knowledge of Paul Nieuwenhuis and Peter Wells (Centre for the Automotive Research, Cardiff University, Wales). Peter and Paul helped me with data collection, interpretation and later reviewing early versions of Chapter 7. Whenever I spoke with Peter and Paul I had the feeling I never knew enough about the industry, but this ended up working on my favor, since I profited from their experience by checking every nut and bolt of this complex sector. More recently, Clovis Zapata also helped me to have a better understanding of the intricacies involved in bio-fuel projects in Brazil and elsewhere. To my dear auto specialists from Wales, *diolch yn fawr!*

ACKNOWLEDGEMENTS

Sweden has also been a very influential place in this journey. Indeed, the early thoughts about the need for a book such as *Sustainability Strategies* came from my interactions with Master and Doctoral candidates at the International Institute for Industrial Environmental Economics (IIIEE), Lund University, in the early 2000s. Teaching the Strategic Environmental Management (SEM) course was anchored by an action-research program, in which, during a period of four years, Michael Backman and I tested the foundations of SEM in 35 real-life reference companies of multiple sizes and sectors. As the reader will notice, some of these organizations are represented in the cases studies presented in various sections of the book. In my frequent visits to Sweden for lectures and seminars, my former students and colleagues from the Institute – Luis Mundaca, Kes McCormik, Murat Mirata, Beatrice Kogg, Thomas Lindhqvist, Naoko Tojo, Philip Peck, Tareq Emtairah, Oksana Mont, Åke Tidell and Katsiaryna Paulavets – provided me with material and feedback on the earlier versions of the chapters. Luis was of special help to me by being more than patient in explaining the intricacies involved in tradable certificates and revise several versions of the section on Emission Trading Schemes (ETS, Chapter 3). The help of Yuliya Makarova with the bibliography and figures was also crucial in getting the book in the right formatting. For all the support from the Institute people, *tack så mycket!*

Francesco Zingales (Nexance, Italy), has been following the developments of the book since the early days and has been instrumental in bringing me down to earth with requests for accessible language, so that managers and MBAs can more easily digest the very complex issue of sustainability management. Francesco, *Grazie Mila!* From Holland, *Dank je wel* also to Frank Den Hond (Free University, Amsterdam) for his insights on strategy and the opportune feedback in some of the chapters.

Reflecting the diversity of the project that culminates in this book, in Brazil I counted with the incredible support of Fernando Von Zuben (Sustainability Director, Tetra Pak) and his very engaged group of collaborators, who received me in a truly Brazilian style, touring me throughout the often tough reality of social inequalities of waste collectors, and a multitude of organizations that form an amazing recycling network. Besides the support during my field work Fernando also provided me feedback in early versions of the chapters. On a more personal level, from Brazil I always counted with the unconditional support of my family. *A todos vocês, muito obrigado!*

Finally, I was very pleased with some recent help received from California. Susan Shaheen (Director, Transportation Sustainability Research Center University of California, Berkeley), provided me relevant data, publications and feedback on car-sharing. Other crucial insights came from Palo Alto via the collaboration with Better Place. Many thanks to Shai Agassi (President and CEO) for sharing some inspiring thoughts, and Guryan Tighe, for facilitating the interviewing process with senior executives of Better Place. I do hope that Shai's entrepreneurship and other examples presented in the book can inspire others to transform the sustainability vision into feasible business strategies and practices.

# PART I
## FUNDAMENTALS

# 1

## WHEN DOES IT PAY TO BE GREEN?

The possibility of business to profit from environmental investments – the *win–win hypothesis*[1] – has captured the imagination of academics, managers and the general public for quite some time. If investing in environmental protection were profitable,[2] normal business practices would be conducive to sustainable societies. Based on this premise, academics have persistently looked for causal relationship between environmental investments and variables such as stock price and market share.[3] The business case for sustainability exists indeed. Business schools around the world teach success stories of environment-oriented investments (or eco-investments, for short) that paid-off, generated competitive advantage or even new market spaces. But if there are so many advantages for business, why is corporate proactive behavior not a widespread phenomenon? Why hasn't commerce yet led us to sustainable societies? Although simple, it took a while for people to realize that the profitability of environmental investments is similar to other issues in business: it is conditional to specific circumstances. As Forest Reinhardt[4] put it, the question is not whether corporations can offset the costs of eco-investments, but when it is possible to do so. In his view, the possibility for corporations to profit from eco-investments depends on "the economic fundamentals of the business, the structure of the industry in which the business operates, its position within that structure, and its organizational capabilities".[5] Hence, directing a firm's efforts toward profit generation from cleaner technologies or green products might make business sense in certain circumstances, not in all.

If certain conditions favor eco-investments to generate returns, can they also become sources of competitive advantage or generate new market spaces? For quite some time, academics and practitioners have also claimed that this is the case.[6] More recently, as sustainability further enters the political and managerial agenda, the claim that eco-investments pay-off became almost dogmatic. According to this logic, a positive correlation between corporate eco-investments and profits is the best motivation for firms to go beyond compliance (i.e., beyond what they are required by law), and industrial competition itself can generate sustainable practices. Unfortunately, empirical examples of such correlation are exceptions rather than the norm. Although opportunities to generate some returns are available to most firms, only a few have been able to enhance their competitiveness based on beyond compliance management.

Overall, *when it pays to be green* remains an open question. Although the number of publications in the topic has greatly expanded in the past decade, material based on solid foundations is still scarce. They are normally based on a myriad of case studies that do not have a frame of reference to establish a relationship between themselves and, more importantly, between practice and theory. Making things murkier is the predominance of normative texts. Paraphrasing Michael Porter, corporate environmentalism "became a religion with too many priests",[7] and too many solutions have been offered based on unsound business rationales. As a result, managers faced with the task of prioritizing eco-investments find themselves in a terrain crammed with dubious prescriptions.

This chapter aims to clarify the question by identifying the broad areas in which win–win scenarios can happen. In order to ground the discussion, the following sections present two empirical cases of eco-investments. They help us to understand the scope of *corporate environmentalism*, a term used throughout this book that encompasses the practices taken by any organization in order to reduce the environmental impact of processes, products and services along the entire life cycle. In this respect, the concept of corporate environmentalism is clearly less ambitious than the most commonly accepted definition of sustainability or sustainable development, presented by the Brundtland Report:[8] "the development that meets the needs of the present without compromising the ability of future generations to meet their own needs". The reason for the perspective adopted here refers to the high levels of difficulty to bring the concept sustainable development to an operational level. Even though ecological sustainability is a desirable outcome for most of us, it is more a societal vision that can inspire academic research and management practice than a model that can be tested or implemented.

4

Nonetheless, by focusing on the areas in which win–win scenarios can emerge, skeptical managers may be convinced to adopt increasingly ambitious environmental strategies and practices. Likewise, Master of Business Administration (MBA) students and academics working in areas such as strategy, operations management and marketing also need sound rationales to be certain about the relevance of environmental issues in business. Together, the practice and theorization of sustainability management will eventually contribute to more sustainable systems of production and consumption. After all, if sustainable development is a vision to be pursued by all, helping the skeptics to identify when and how they can contribute is crucial. The case studies presented in the next sections open this argument. Together, they will suggest the opportunities and limitations of corporate environmentalism and the basic conditions for the identification of *when it pays to be green.*

## GREENING AS A COMMITMENT[9]

Founded in 1951 as one of the first packaging companies for liquid milk in the world, Tetra Pak evolved to a global supplier of packaging systems for liquid food products. Today, the company operates in 165 countries with over 20,000 employees, providing integrated processing, packaging and distribution line and plant solutions for foods manufacturing. Tetra Pak has long been committed to running its business in a sustainable manner by setting goals for continuous improvement in development, sourcing, manufacturing and transportation activities. As one of its policies, all Tetra Pak packages have to be suitable for recycling. The company also supports customers to finding solutions for their packaging material waste, and is committed to facilitate and promote local collection and recycling activities for post-consumer carton packages. Tetra Pak also endorses principles in the areas of human rights, labor and the environment via organizations, such as: the United Nations Global Compact;[10] NetAid, a growing network of people and organizations committed to ending extreme poverty;[11] and the International Business Leaders Forum (IBLF),[12] a not-for-profit organization that promotes responsible business practices.

The pioneering use of aseptic technology for packaging in the 1960s by Tetra Pak represented a breakthrough in the liquid-packaging industry. Aseptic technology keeps food safe and fresh by maintaining flavor for at least six months without refrigeration or preservatives. The aseptic packaging, better known as long-life packaging, is made of six layers of three different materials: long-fiber duplex paper (75 percent by weight), low-density polyethylene (20 percent) and aluminum (5 percent). Until recently, it would be difficult to identify any problem with the aseptic technology. But the multi-layered packaging material makes the total recycling process very difficult.

The Brazilian subsidiary of Tetra Pak started its operation in 1957 and by 2005 it was the second largest operation of the group, surpassing 8 billion packages in sales.[13] As an extension of the Swedish culture, the Brazilian branch has consistently invested in environmental protection.[14] When Mr Fernando Von Zuben joined Tetra Pak in 1995 to be the head of the environmental department, he understood that his major challenge would be to close the cycle of post-consumption aseptic cartons. The first major barrier was the low levels of selective collection of household waste. Tetra Pak addressed this problem by sponsoring environmental education. Since 1997, the company has been distributing a pedagogical kit developed by the University of Campinas, in the State of São Paulo, to support classroom discussion about the problem of urban solid waste, the importance of selective collection, and the environmental, social and economic benefits of recycling. By 2005 around 2200 primary school teachers were trained to use the instructional material, which have been distributed to more than 5 million students in 40,000 schools around the country. In 2006 Tetra Pak made the program available via videoconference.

Another form of increasing recycling rates of aseptic packaging is to provide local councils and cooperatives of collectors with technical expertise. Between 1997 and 2005, more than 200 Councils were offered Tetra Pak's expertise in recycling. Normally, Tetra Pak motivates councils to establish stable and reliable collection systems via cooperatives of collectors. The cooperatives provide a more systematic approach to the problem of collection but their formation serves a much broader social aim. Members receive a closer social assistance via education and medical care, among others services. In its efforts to support the formation of cooperatives, during 1997–2005, Tetra Pak sent more than 4 million brochures to municipalities and cooperatives of collectors. In order to increase process efficiency, during that period Tetra Pak donated 30 press machines to cooperatives in various states of Brazil. Such support resulted in aseptic cartons representing between 6 and 10 percent of the income of collectors by 2005.

Once the material is collected, then the proper recycling could start. Until 1997, however, several technical and commercial hurdles limited recycling. The recycling of paper content in aseptic packaging has never represented a major problem. It can be extracted from the sandwiched layers of polyethylene/aluminum (Pe/Al) via hydrapulper – standard equipment in paper recycling plants, which have to be adapted for the separation of Pe/Al – the filtering of eventual residues. Until 1997, such requirements made paper manufacturers reluctant to use the fiber from aseptic cartons. Tetra Pak addressed the problem by showing manufacturers the technical advantages of the fibers from aseptic packaging. The fibers are new and, therefore, longer than those that have already gone through a recycling processes, what convinced, recyclers of paper to gradually start giving priority to Tetra Pak aseptic packaging. The range of recyclers is very broad, varying from medium and small local producers to Klabin,[15] a world-class corporation specialized in the manufacturing of paper.

After the extraction of paper fibers from the packaging, it is necessary to find use for the blend of Pe/Al. Until 1999, there were no applications for the Pe/Al coming either directly from the Tetra Pak factory or as post-consumption waste. The environmental team of Tetra Pak was given the challenge to find solutions. After a few years of trial and error, the blend Pe/Al was used for the production of roof tiles and boards. Basically, the process consists of pressing the material under temperatures around 180°C, with subsequent cooling, acquiring the shape of plain boards or corrugated roof tiles. Even with very low levels of process optimization, the business is profitable. The initial investment in equipment is about US$58,000 and revenues of US$45,000 per year are feasible.[16] Profitability depends on various aspects. Location strongly influences the costs of transport and the price for the raw material. For, instance, in 2005, out of the ten producers of roof tiles and boards in Southern Brazil, the ones operating in Paraná and São Paulo states were in the best position. Since profit margins are relatively low, fluctuations in the price of raw material can have a perverse effect on the business viability.

The Pe/Al is also used as raw material in plastic products. After the Pe/Al mix is taken out from the hydrapulper as rejects of the process of paper recycling, it is cleaned to remove fiber dried up and, in order to increase the homogeneity of the material, grinded. After this, the rejects are fed into an extruder, which will melt them down at temperatures of 200°C and then extruded into Pe/Al pellets. These pellets are sold to manufacturers of plastic materials, such as buckets, brushes and handles for tools. The Pe/Al pellets are used as substitutes for pure plastic material (polyolefin, normally polypropylene or polyethylene).

In this case, even though the aluminum content is a contaminant of the plastic material, because it does not compromise the product performance, it can be used as a substitute for pure plastic.

Even though Tetra Pak's initiatives in Brazil resulted in greater rates of collection and recycling of aseptic packaging, until 2004 the solution was still an incomplete one. Because there were no technical ways of separating the layers of polyethylene from aluminum, they were used for applications that require lower performance than in aseptic packaging, such as brushes, brooms, boards and tiles. Even though this is a better solution than land filling, it still does not close the material cycle. After more than ten years of research and development and the collaboration between Mr von Zuben of Tetra Pak and Dr Roberto Szente, the head of the plasma research team of IPT,[17] a Brazilian research institute and an expert in plasma technology. Closing the cycle of aseptic cartons became possible via the use of thermal plasma technology. The process comes as a new recycling option, which separates the three package components allowing them to return to the production chain as raw materials.

Metallurgical companies have traditionally used plasma technology for metal recovery. The use of plasma for the recycling of aseptic packaging is, however, novel. Plasma technology is highly efficient: 90 percent of the energy yield is actually achieved in the process. In comparison, doing the same process with natural gas, the efficiency would be down to 25–30 percent; the aluminum would be contaminated and the plastic would burn. Besides the eco-efficiency of the process, emissions during the materials recovery are near zero because the atmosphere of the reactor has close to zero concentration of oxygen.[18]

After running the pilot plant for 24 months, the quality of the plant output (paraffin wax and aluminum) was so encouraging that a proposal for a joint venture emerged. In 2004, Tetra Pak, TSL, Klabin and Alcoa (an American producer of aluminum in Brazil) constituted a new firm named Edging Environmental Technology. The Thermal Plasma plant installed in the city of Piracicaba (in the State of São Paulo), has the capacity to process 8000 tons of Pe/Al material per annum (equivalent to 32,000 tons of aseptic packages), representing 20 percent of all Tetra Pak packaging produced in Brazil (160,000 ton/year). The plant required an investment of US$5 million, shared among the four partners. During 2004–2008 the plant operated at 50 percent of its potential, reaching full capacity in 2009 with the installation of the second reactor.

Alcoa, the supplier of the aluminum foil for aseptic packaging, has been buying the recycled aluminum for the production of new foils. Aluminum has been sold to Alcoa at 95 percent of London Metal Exchange (LME) price. The paraffin wax has been sold to the national chemical industry to be used as wax emulsion for paper. The sales of paraffin and aluminum produced by the plant are expected to result in a payback period for an investment of two years only. Such investment prospects lead to an increased interest outside Brazil. Plasma plants are planned for Spain and Belgium. In October 2005, The Brazilian Industry Council, a governmental organization, awarded Tetra Pak a prize in the category of sustainable development for the plasma plant factory. The project has been so successful that National Geographic Channel included it in its documentary series about *Megacities*.[19]

Overall, Tetra Pak efforts to close the aseptic carton cycle resulted in great increases in recycling rates. Such results are important not just because the overall environmental impact has been reduced, but also because the recycling networks promoted by the company have generated new sources of wealth and alleviated poverty. The implementation of the Plasma plant represented an opportunity for Tetra Pak to obtain a return from the US$1.2 million per year invested since 1997 to close the aseptic carton cycle. Curiously, the company was not interested in such type of return on investment. According to Mr Fernando von Zuben, Tetra Pak does not have the intention to profit from the plasma technology. The efforts of the company to close the material cycle are part of the overall philosophy of the group toward excellence in corporate environmentalism.

## GREENING AS A CORE COMPETENCE[20]

In 1991, Bakelittfabrikken (BF, for short), a large Norwegian firm with considerable know-how in thermoplastic molding, led the consortium to establish The Personal Independent Vehicle Company (PIVCO). The new company intended to develop and produce a unique vehicle to capture part of the niche market for urban and suburban eco-friendly transportation. The Electric Vehicle (EV), initially called *City-Bee*, and later renamed *Think*,[21] was expected to be the second household car, as well as a vehicle that could be used to distribute goods and services by organizations, such as energy utilities, municipalities with services for the elderly, telecommunications companies and car rental firms.

Apart from the in-house development of the body panels and aluminum space-frame, PIVCO chose to purchase the vast majority of parts from established suppliers in the auto industry. Overall, about 70 suppliers provided the components for the *Think*. This was less than half the number of suppliers of Internal Combustion Engine (ICE) cars, suggesting the reduced number of components used by the vehicle. Several innovations were associated with the *Think*. Unique design and manufacturing concepts entailed a minimal environmental impact, as compared to the production systems of traditional car manufacture. The use of aluminum in the space-frame, and the thermoplastic in the body made the *Think* a distinctive vehicle. The space-frame is mounted on a folded-welded steel platform. In this way, the design concept of the platform (or the floor of the car) bypassed the high investments in press shops, characteristic of traditional floorpans. (Chapter 7 further discusses this point).

The body is produced with the same industrial process formerly used by BF to manufacture boats. Color is added to the thermoplastic during the molding process of the body parts, thereby making painting unnecessary. Such innovation bypasses one of the most polluting processes in traditional car assemblies, as it eliminates the need of a paint shop. In addition, it significantly lowers the capital necessary to set up the assembly plant. Although the thermoplastic body panels may not have the smooth finish of painted panels of traditional steel bodies, they do offer an advantage that, when dented, they do not loose color, as would normally happen with painted panels. Further to this, the thermoplastic panels are both rustproof and recyclable; and the overall modular approach used by PIVCO facilitates the disassembly and recycling of the material.

The production paradigm guiding PIVCO also differed substantially from that of the traditional (and current) practices in auto manufacture. Volumes of 5000 units per year should be sufficient to reach the break-even point[22] – an impressively low volume when compared to those of the more traditional car manufacturers. For instance, the two-seat *Smart ForTwo* car, launched by Mercedes-Benz in 1997 (and discussed in more detail in Chapter 7), requires a minimum volume of around 100,000 units per year to pay for the depreciation on start-up costs. This means that the break-even point of the *Think* represents a mere 5 percent of that of the *Smart*.

The production system also entails a novel approach in retailing the vehicle. Since the early days, the company intended to focus on the know-how in EV development and manufacturing principles, and establish joint ventures to produce the *Think* close to the target market – mirroring, in some respects, the principle used in franchising. Similar to principles used in the installment of the McDonald's eateries, the general production principles and management techniques for the *Think* could be applied to other parts of the world, where the demand for the vehicle could justify the investment. Factories could be established on production (and expansion) volumes of 5000 units per year.

While the innovative approach in the design and manufacturing of the *Think* reduced installment costs, such savings were countervailed by the high costs of the nickel-cadmium battery pack used in the vehicle. The batteries had the advantage of a useful lifespan of approximately 200,000 km, or the equivalent of 20-years' usage, calculated at 10,000 km per year. However, the 247 kg weight of the battery pack corresponded not only to roughly one third of the total weight of the *Think* (917 kg unloaded), but also in terms of the final price of the vehicle. In October 2000, the retail price for the *Think* in Norway was €25,000, of which €7500 was due to the battery pack. For consumers, the high initial purchasing price could be mitigated throughout the life cycle of the vehicle, but only if they kept the car for 20 years.

The relatively low-capital injection necessary to establish the enterprise, as compared to similar investments undertaken in the auto industry, was not sufficient to prevent financial trouble for the company. Between 1991 and 1998, PIVCO relied on its consortium members and the Norwegian government to invest approximately €35 million in the project. However, the enterprise faced an enormous challenge when an additional €9 million was needed to assist in the recruitment of workforce and in the purchase of components for production of 5000 commercial vehicles per annum. The combination of the stock market crisis in 1998 and the refusal of the Norwegian government to continue investing in the company resulted in PIVCO going insolvent. In October 1998, the financial crisis forced the factory close down and the declaration of bankruptcy. Two weeks later, the company was bought back by the PIVCO management team, BF and its employees.

That was not the end. In early 1999, PIVCO was rescued by Ford Motor Company and became a central player for Ford's plans to comply with the zero emissions mandate of California. The designed regulation required at that time that, by 2003, car manufacturers would have to sell a certain percentage (depending on past sales) of their car fleet as zero emissions vehicles. In practice, this meant EVs. In January 1999 Ford bought 51 percent of PIVCO, and later the remaining 49 percent of the renamed Think Nordic AS, which led the development of environmentally friendly vehicle technology in Ford. Late in the same year Ford started the production of the *Think City* (called A266 by Ford).

By March 2002 Ford had produced 1005 vehicles, making the *Think* one of the largest fleets of EVs on the road. The car was sold in 14 countries, but the main clients were in Norway, Denmark, Sweden, as well as some selected cities in Europe and the United States. Ford also invested in the development of a new generation of the *Think City* (A306), which aimed to better meet the needs of the American market. The factory was prepared to produce the new car and launch was planned for the fall of 2002. In that August, however, Ford announced that they were pulling out of Think Nordic, stating that they preferred to concentrate on other alternative technologies such as hybrids and fuel cells.

On February 2003, Think Nordic was bought by Kamkorp Microelectronics, which had interests in the field of innovative transportation. The business focus of Think Nordic – on homologation and commercialization of EVs – compliments Kamkorp's competences in drivetrains and microelectronics. Shortly after the takeover, work commenced on the *Think Public*, a vehicle designed for the specific use in urban centers and closed areas. In early 2006, Think Nordic[23] changed hands once again. The investment group InSpire acquired all its assets. The group relaunched the new *Think City* model (A306), which Ford developed whilst they owned the company. At that time, battery technology had not evolved enough such that the range of the vehicle was compromised. The new investor group has access to new and advanced technology, which will be further developed and utilized to improve the vehicle with particular emphasis on performance and driving range.

The new owners view the factory in Norway as a base for technology development, and exploit partnership and license production opportunities around the world. The group has invested in renewable energy, fuel cells and other technologies, which will have significance for *Think*. The high fuel prices and a growing concern over the effect of conventional transport on climate change have once more revamped the interest for eco-friendly vehicles. Think Nordic has products that address these issues but, considering its bumpy history, will it ever realize its business potential?

## THE FRONTIERS OF CORPORATE ENVIRONMENTALISM

The relevance of environmental issues in business is the obvious common element of the cases presented in the preceding sections. Together, the cases suggest that an extensive array of actions to reduce the environmental impact of processes and products is available to firms. As a whole, these compendia of activities define the scope of corporate environmentalism.

Figure 1.1 depicts the scope of corporate environmentalism in the form of a graph. The vertical axis describes the actions companies may take toward environmental protection or restoration, which will eventually generate *public benefits*. Purified water resulting from recycling processes in a factory re-sent to a river is a clear example of a public benefit. The horizontal axis represents actions that generate *private profits*, such as cost savings for the firm resulting from reduced water consumption. Actions that generate more public benefits than private profits are closer to the top border of the cone, defined by Line E. The opposite situation is depicted by Line B, in the lower part of the cone. The area between Lines E and B define the scope of corporate environmentalism with potential to generate both public benefits and private profits – that is, the win–win scope of sustainability strategies.

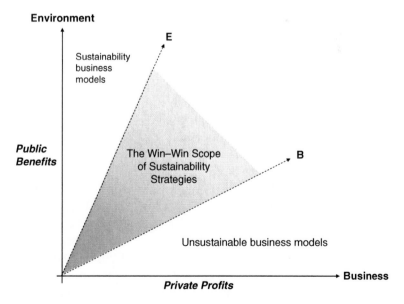

FIGURE 1.1 **The Scope of Corporate Environmentalism**[24]

In generic terms, it pays to be green when environmental investments are located within the cone of Figure 1.1. The closer eco-investments are to the right-hand side of the horizontal axis (Business), the more lucrative they are (maximum private profits). On the other hand, eco-investments situated closer to the top end of the figure, in the vertical axis (Environment), are more environmentally sustainable (maximum public benefits).

In principle, a company could work very close to the vertical axis (Environment) hence, outside the area of win–win scope. But in doing so, it risks not to fulfill shareholders' expectations – that the company should work toward economic value creation. On the other hand, if the company only seeks business opportunities without considering the impact on the environment, it would be working toward business as usual, as it has been popularized, described in the horizontal axis. Although in many parts of the world companies can still work as pure business, sooner or later they will eventually face opposition from environmentalists.

Corporations should consider shareholders' expectations of economic value creation but it does not mean that they should not try to push the upper boundary (Line E) as far as possible. After all, innovations can generate returns from areas that were formerly considered unprofitable, as the plasma plant for the recycling of aseptic cartons suggests in the Tetra Pak case. In fact, from the 1990s onwards, several models or frameworks were developed for businesses that are keen to push these boundaries forward. These models normally adopt the ecologist perspective: they depart from the constraints faced by the planet (air pollution, water scarcity, loss of biodiversity, etc.), and prescribe what business should do to reduce its environmental impact, which would eventually be conducive to sustainable societies. In other words, these are sustainability-oriented frameworks that can guide firms to be more ecologically sustainable.[25] What managers should not expect, however, is that such actions will always create economic value.

Tetra Pak and Think Nordic are examples of companies that have been trying to cover the wider possible scope of corporate environmentalism, represented by the win–win zone. However, as the cases clearly show, the outcomes of their efforts are very distinct. The Tetra Pak case suggests that in some circumstances corporate environmentalism can be driven mostly by managerial commitment. As it has been the case of many leading firms around the world, moral and ethical commitments may assume more significance in shaping organizational culture and attitudes than market forces, and eco-oriented investments in processes or products do not seek justification simply through recourse to economic principles. For more than a decade, Tetra Pak Brazil has invested in *product stewardship* – the minimization of pollution not only from manufacturing but also from all environmental impacts associated with the full life cycle of a product[26] – without any indication that the investment would generate economic returns. And even when returns became possible with the creation of new business for the recycling of materials, the company refrained from appropriating them.

Of course, a more cynical view of Tetra Pak's altruistic policy would consider the indirect influence that the efforts toward product stewardship have in increasing the market value of the company. After all, as Chapters 4 and 5 explore in more detail, reputational value is the essence of intangibles, which have became increasingly important for companies. The achievements in Brazil and in many other countries certainly increased the reputation of Tetra Pak as a leader in social and environmental responsibility. After all, as Brown and Dacin[27] asserted, "what consumers know about a company can influence their beliefs and attitudes toward products manufactured by that company". Being an exemplary corporate citizen, as the legion of Corporate Social Responsibility (CSR) proponents have stated, can positively influence the value of firms. Moreover, by practicing product stewardship, Tetra Pak also prevents eventual risks associated with legislation based on Extended Producer Responsibility (EPR), which has increasingly being implemented in the European Union.[28] Hence, even though the pro-activism of Tetra Pak may be motivated by ethical commitments, they are certainly in line with shareholders' interest.

Figure 1.2 shows a line dividing the cases. Tetra Pak's actions toward product stewardship are to be found somewhere in Zone E, characterized by the *non-rival* or *non-competitive* nature of environmental strategies. As it happened during the past years, the investments in education and the support Tetra Pak has given to local councils and cooperatives of collectors resulted in increasing levels of recycling (with consequent lower impact of aseptic packaging in the environment) and the creating of jobs and better social conditions for thousands of people (the empowerment of the *base of the pyramid*[29]). In other words, clear public benefits (vertical axis). From a strict economic sense, the investments did not result in direct tangible *private profits* for Tetra Pak (horizontal axis).

The non-rival aspect also means that the several actions aiming to increase the recycling rates of aseptic packaging (the eco-investments)

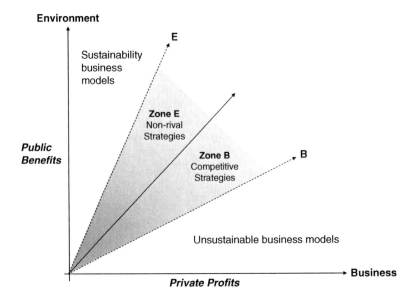

FIGURE 1.2 **Competitive and Non-competitive Strategies**[30]

do not compete with similar actions taken by other companies operating in the same industry. On the contrary: if competitors do as Tetra Pak does, the levels of collection will increase, resulting in higher levels of recycling rates for aseptic packaging. In other words, the main result is a clear *public benefit*. The clients of the packaging equipment manufactured by Tetra Pak, as well as the consumers of milk, juice and other liquid products that use the aseptic packaging pay little attention to the fact that recycling rates have increased. As Chapter 6 explores in detail, they tend to focus on more immediate issues, such as product convenience, quality and price. Such indifference of clients and consumers to the product stewardship efforts makes the recycling of aseptic cartons a non-rival business issue. In such cases, if other firms are not willing to adopt beyond compliance practices, as Tetra Pak does, increasing recycling rates are possible only via enforced regulation.

The case of Think Nordic shows a different side of corporate environmentalism: the vital importance of obtaining returns from eco-investments that present a *competitive* nature. Although the very foundation of the enterprise was based on a good degree of environmental commitment, the business passed through several critical periods simply because economic returns were difficult to obtain from the core business of the company: the electric car (which is by its very nature an eco-oriented investment). Since the main marketing appeal of an EV is its eco-friendliness,[31] for Think Nordic, corporate environmentalism is closely coupled with the core competences of the business. Both the business model and the general corporate strategy relate directly to the environmental strategy of the enterprise. Even though the car produced by Think Nordic is electric, it competes not only in the niche of electric cars but also with conventional cars powered by Internal Combustion Engines (ICEs). The Think competes, for instance, with other two seat model cars, such as the *Smart ForTwo* (owned by Mercedes), or even with small four seats ICE cars, such as Citroen C2, Fiat Punto and Renault Clio. In this respect, the environmental advantages of EVs face direct competition from motoring attributes of conventional cars, such as speed, range and road performance (Chapter 7 readdresses this point). For this reason, the environmental strategy is, in fact, a *competitive strategy*, defined by Zone B.

Competitive strategies – environmental or otherwise – can be seen as *zero-sum game*, in which one company's gain is achieved at another company's loss. As it has historically been the case, sales of conventional cars reduce the market size for electrics in the same segment. This might sound like bad news for eco-friendly products. But environmental prerogatives seem to be growing in importance to consumers. In any case, the most important aspect here is not whether green processes, products or services will eventually outdo conventional ones, but the understanding that eco-investments have higher chances to pay off when their attributes make them an asset subject to competition. In the perspective of Figure 1.2, competitive environmental strategies will emerge when the chances of creating private profits from eco-innovations are higher than the public benefits resulting from them.

## METHODOLOGICAL CHALLENGES IN *WHEN IT PAYS TO BE GREEN*

Defining the scope of win–win scenarios and the broad conditions in which companies can purse them, as it was done in the previous section, is the first step toward answering *when it pays to be green*. Unfortunately, the move from a generic understanding to particular circumstances encompasses several methodological challenges. This is because, besides the multiple dimensions, timings and contexts comprised by the question, the answers serve two distinct audiences differently. For academics, being green is mainly a *comparative* issue. Scholars are normally interested in identifying the greenest company of a sample, according to methodologies that permit the elaboration of reliable correlations.[32] Managers, on the other hand, are interested in knowing the return resulting from specific eco-investments, which can eventually enhance their competitiveness. In other words, for managers, being green is a *competitive* issue.

There are points of convergence between academics and executives but they depend on the methodology used in the studies. Being green often means different things to different people, researchers in particular. Traditionally, greening has been associated to overall environmental performance of industrial plants, and the accumulated emissions of factories have often been the main variable used to measure it. This is surely a valid approach but, in this case, environmental impacts associated with the phases of use and end-of-life of products do not count. Automobile assembling plants, for instance, have significantly reduced the emissions of production processes. Such reductions however, count only for around 10 percent of overall environmental impact of cars during their life cycle. Hence, in order to identify the sustainability leader in the industry, calculations have to consider not only the emissions of factories but also those associated with car use, which represent more than 80 percent of the total impact, as well as the end-of-life impact of vehicles.

Methodological barriers have not deterred academics from doing their best to identify the outcomes of eco-investments. Scholars searched for correlations between accumulated emissions of factories[33] and financial indicators, such as return on assets and the stock price.[34] They also looked for correlations and causality between environmental performance and competitive advantage by showing, for instance, that some firms increased their market share when they reduced toxic emissions. In comparing the environmental performance of the sampled firms with the performance in the stock market or other financial indicators, these studies assume a macro-comparative perspective. Such studies are fundamental to advance the knowledge in the field of sustainability strategies. However, as the corporate scandals of Enron in the US and Parmalat in Italy suggest, the real value of companies can be subjected to accounting manipulations, distorting correlations between financial and environmental performance. Even if accounting systems provide reliable data, speculative price fluctuation of stocks is just one of the intervenient variables rendering difficulty in establishing reliable correlations. Additionally, stock prices of firms supplying consumer markets tend to be more sensitive to variations in the attributes of the products (oil, for instance) than on the environmental performance of industrial processes (refineries).

An additional problem with the comparative studies is the bias toward large firms due to data availability. Access to reliable data influences most studies to focus on large, publicly listed companies. Although many sustainability leaders are indeed large firms, not always they are the ones presenting outstanding environmental performance. In Sweden, a great number of Small and Medium Enterprises (SMEs) have extremely high standards of environmental performance, but because they are small, very seldom they are included in comparative studies. Overall, it is extremely difficult to identify *when it pays to be green* from a macro-perspective.

As the case studies discussed at the beginning of the chapter suggest, being green is a broad and complex issue, encompassing all stages of the life cycle of products. Eco-investments include innovations in design, such as the color-injected molding for body parts of the *Think* car, which resulted in the elimination of the paint shop, and consequent lower emissions of the factory. In most cases, however, the range of eco-investments is far broader than the factory walls. Since the *Think* is an EV, its most important eco-oriented investment is its zero emissions during use, but the easiness of disassembling its parts for re-use or recycling can also be considered eco-design investments. The time spent to design parts that are easily disassembled can eventually generate economic returns at the end-of-life of the vehicle.

In fact, the life cycle perspective of *Think* and the product stewardship practices of Tetra Pak bring us to another methodological issue involved in the return of eco-investments. The timeframe for the analysis of these investments is somewhat controversial. Many eco-investments may present long lead times before being translated into performance. Training for the implementation of an Environmental Management System (EMS) is a typical example; it may not lead to better environmental performances in the short term, but it may be crucial for the future performance of the firm.[35] In this case, what should be the best period for the measurement of the return of an EMS implementation? Product-related costs with Life Cycle Assessment (LCA), eco-labeling and market development constitute another type of investments that take long time to mature, and any short-term evaluation would be misleading. The payback period for an eco-brand might be five or more years, so any short-term analysis would fail to recognize the value of the investment.

As it will be explored in detail throughout the book, the context favoring greening also influences the chances of an eco-investment to pay-off or generate competitive advantage. The structure of the industry, encompassing degree of industrial rivalry, as well as the political economic context in which the business is embedded can significantly influence the chances of an investment to succeed or not. For instance, the profitability of an ecological brand to succeed depends on the cultural trait of customers, reflected in their degree of environmental awareness and willingness to pay more for greener products (Chapter 5 addresses these issues in more detail).

Finally, the methodological challenges do not end in the value of tangible assets. The return of green investments very often occurs in the form of intangibles. As many of the cases presented in this book show, intangibles increasingly represent an important share of the return on eco-oriented investments. When evaluating the success of corporate environmentalism, the creation of new competences or, as the Tetra Pak case indicates, enhanced corporate reputation, cannot be overseen. Overall, the identification of circumstances rendering environmental investments profitable simply cannot be based solely on quantitative correlations between empirical variables. The methodological challenges discussed in this section suggest that, in order to identify *when it pays to be green*, we need to decouple the terms of the question and consider:

- When: a clear timeframe, and the context in which the company operates;
- Pays: quantitative and qualitative data, as well as the tangible and intangible value created by the eco-investment;
- Green: a clear definition of the type of eco-investment.

Such considerations point toward the case method, a research approach that focuses on understanding the dynamics present within single settings. The main strength of case studies is the likelihood of generating novel and empirically valid theory.[36] On the other hand, the extensive use of empirical evidence can yield theory that is overly complex, narrow or idiosyncratic. In order to curb such limitations, the analysis of multiple case studies (practice)[37] can be used to generate an ampler, grounded and less-idiosyncratic frame of reference (theory). As it has been done in the study that resulted in bringing out this book, a number of cases were jointly studied to inquire into the profitability of eco-investments. In many instances, the collection of data via case method was also combined with action-research: a process by which researchers work with members of an organization on a matter of importance to them, and in which they intend to take action based on the results.[38] Although other methods are certainly possible to be deployed, direct access to the organizational reality helps identifying the subtleties involved in corporate environmentalism. By combining multiple case study and action-research, the study expressed in this book[39] does not eliminate the methodological limitations involved in this type of research. Nonetheless, all considerations highlighted in the bullet points have been taken on board in the process of identifying *when pays to be green*.

## CONCLUSION

This introductory chapter showed that, while the scope for corporate environmentalism is vast, only a few actions toward environmental protection will generate economic returns or competitive advantages. Although simple, the reason for this is seldom recognized. The scope for non-competitive actions is much broader for those in which the gain of a business is achieved at another business' loss (the *zero-sum* game). In other words, for most eco-investments to generate market advantages, they need to compete with alternatives – green or otherwise – in the marketplace. Some eco-investments will only generate return when internal or external stakeholders attribute value to them, which often is intangible. Taken as a whole, the first step toward answering *when it pays to be green* requires us to focus on the competitive nature of eco-investments.

Some skeptics may question the usefulness of competitive rationales for sustainability management, as I propose in this book. They may argue that, as society evolves and business assumes more environmental and social responsibilities, there will be no scope for a perspective focusing on the competitive aspects of environmental management. Without making any judgment whether it is good or bad for society to be more or less competitive, such argument does not pass a simple test. If in a near future all corporations in the world took social and environmental responsibility on board, they would be doing as much as they could to reduce their impact on nature and increase social welfare. A sustainable society would eventually emerge from these practices. Would this mean the end of rivalry? In particular, would it mean that environmental or social aspects could not be used as means of cutting costs, differentiators or vehicles to create new market spaces? Although we can imagine that it would be more difficult to do so, rivalry could move to a different level but would certainly not disappear. As long as commerce exists, competition will drive it.

Others, such as Kim and Mauborgne, the authors of *Blue Ocean Strategy*,[40] would even argue that companies should simply avoid competition by creating new market spaces. Why should companies fight for existing markets if they can create new ones? As Chapter 7 explores in detail, this is certainly true for players that have enough innovative capabilities, resources and, very often, the guts to reinvent markets. These companies will certainly be the winners of the sustainability revolution. But for the majority of firms, such as suppliers of industrial markets, the scope for radical strategic moves is more compact than executives managing those businesses would wish for. The next chapters will explore these limitations in more detail and provide guidance not just to the companies that can afford deploying radically new strategies but also for the vast majority of corporations that need to streamline their sustainability strategies in order to operate in existing markets.

# 2

# WHAT ARE SUSTAINABILITY
# STRATEGIES?

Organizations will always have room to improve the productivity of resources and so reduce their overall environmental impact. Indeed, in trying to go beyond compliance, firms have increasingly adopted wide spectra of social and environmental prerequisites into their strategies and practices. But while most companies are expected to become better citizens, in each industry only a few will be able to transform environmental investments into sources of competitive advantage. As any other issue in business, environmental management is contingent to the internal capabilities and the context in which firms operate. Considering that sustainability strategies are constituents of the generic corporate strategy, one could then ask: what sustainability strategies are conducive to the creation of competitive advantage or new market spaces?

This book is not so ambitious as to propose a definitive answer for this question. Nonetheless, it frames the question in a manner that facilitates the analysis of circumstances in which eco-investments can eventually payoff in several forms. As the sections within this chapter will explore, the identification of sources of competitive advantage requires us to make a distinction between products/services and organizational processes. This peculiarity is crucial to unfold the intricacies involved in eco-oriented competitiveness. Such distinction requires an approximation between the two leading theoretical approaches in strategic management: Michael Porter's Positioning School (PS) and the Resource-Based View (RBV) of the firm.[1] By showing the distinctiveness of four strategic choices in the area of competitive environmental management, the chapter shows why an approximation between two apparent irreconcilable perspectives can advance the practice and theorization of strategic management. In addition, the chapter also delves into the direction of Blue Ocean Strategy (BOS), proposed by W. Chan Kim and Renée Mauborgne, which represents a rupture from the logic of competitive strategy. In order to adapt these schools of thought for sustainability management, it is first necessary to have a clear idea of what sustainability strategies are, and what they are not.

## WHAT SUSTAINABILITY STRATEGIES ARE NOT

During the 1990s, environmental or sustainability strategies gradually grew in importance to become the theme of many academic and practitioners' publications, as well as the launch of specialized journals.[2] As mentioned in Chapter 1, while academics were interested in identifying factors influencing proactive corporate environmentalism, managers wanted to know how they could transform eco-investments into sources of competitive advantage. The literature mainly promoted environmental management – an area previously delegated to specialists in health and safety – to the status of strategy. Environmental issues became increasingly strategic for firms. What was not clear, however, was how companies could manage such issues, so advantages could emerge. The publication of the book *Down to Earth* in 2000 represented a great step forward in this direction. By applying business principles to environmental management in a systematic manner, Forest Reinhardt[3] emphasized the contingent nature of environmental investments, putting the *pays to be green* debate in the right direction.

Being in the right direction was a good start toward more grounded corporate environmentalism. Reinhardt made much clearer how environmental demands on firms could become sources of cost reduction, pointing out the conditions favoring potential savings in waste and energy within the organization. He also presented a sharp analysis of how the management of risk and uncertainty could become a strategic advantage for proactive companies. Mirroring Reinhardt, Daniel Esty and Andrew Winston in their 2006 *Green to Gold* book considered the management of the Downside aspect of business (reduce risks and costs)[4] one of the two broad categories of environmental strategies; the other being the Upside (increase revenues and intangibles). The authors presented convincing rationales for proactive management of costs and risks, as well as a plethora of useful examples of companies that increased their competitiveness by adopting them.

Reducing costs within the firm has, in fact, been a common prescription in the specialized literature. Practically all textbooks on the subject mention cost reduction and the management of risk and uncertainty as strategies to be pursued by firms. These authors are certainly right in pointing out the importance of reducing environment-related risks and the need to eliminate every source of inefficiency within the firm, as well as – when possible – throughout the entire value chain. But, if such measures are important for the success of any business, why should they be considered strategies? If strategy involves choosing one path over another, why the efficient management of costs and risks would constitute a sustainability strategy? In order to clarify this question we have to recognize that there is a major difference between what constitutes a strategic issue (important) and a clearly defined strategy (choice). Reducing internal costs and proactively managing risk and uncertainty are definitely of strategic importance for firms – the reason *all* firms should do it. But rather than constituting sustainability strategies, cost reduction and the management of risk and uncertainty are, in effect, rationales *for* strategy. They are constituents of what Michael Porter calls operational effectiveness:[5]

> Operational effectiveness and strategy are both essential to superior performance, which, after all, is the primary goal of any enterprise. But they work in very different ways (...) Operational effectiveness means performing similar activities better than rivals perform them. Operational effectiveness includes but is not limited to efficiency. It refers to any number of practices that allow a company to better utilize its inputs by, for example, reducing defects in products or developing better products faster. In contrast, strategic positioning means performing different activities from rivals' or performing similar activities in different ways.

There is no doubt that guidelines for the efficient management of costs, risk and uncertainty, such as the ones presented in *Down to Earth* and *Green to Gold* have great value for business. By adopting the systematic approaches presented in these books, managers will surely be better prepared for their jobs. But the efficient management of costs and risks should not be confounded with strategy, which requires companies "to be better by being different".[6] Fundamentally, strategy entails choice, and choice involves trade-offs. Trade-offs arise, for instance, from inconsistencies in image and reputation, when a company tries to deliver two different kinds of value at the same time, or when trying to be all things to all customers, confusing both employees and customers.

W. Chan Kim and Renée Mauborgne go further suggesting companies to offer a choice so unique to customers that it would allow them to bypass competition altogether.[7] In order to offer uniqueness to consumers, companies need to choose between clear trade-offs. They will need to eliminate, reduce and raise "the factors the industry competes on in products, services and delivery, and what customers receive from competitive offerings on the market".[8] They will also need to create new factors that will allow the company to offer differentiated products and services at low prices – what they termed value innovation – and even be comfortable with the idea of letting some groups of customers go. In other words, in order to have a consistent and clear strategy, companies will have to make tough choices (Chapter 7 explores this proposition in more detail, and adapts it to the realms of sustainability strategies).

What would then be the trade-off involved in cutting costs and managing risks? What would be the alternative? Could any company afford not taking care of such areas and remain competitive? If the short term may mask the answer, it becomes sharply clear in the end. The management of environment-related costs and risks are part of the operational effectiveness of any company and simply cannot be avoided. Although industries present different levels of efficiency and outcomes, all companies have to perform such activities. Eventually, managing costs and risks will become just a licence to operate. The same can be said about the Upside aspect, proposed by Esty and Winston. Increasing revenues and managing intangibles is not a choice but a business imperative. Esty and Winston are right in pointing out that such tasks are strategically important for the success of companies, and providing managers with tools for the efficient management of risks and costs is definitely useful. But the fact that such activities are important for business do not automatically transform then into sustainability strategies. For that, we need more. If strategy is "doing better by being different", managers need to identify how companies can be different by making specific choices about eco-investments. The debate within the field of strategic management can help in clarifying this point.

## COMPETITIVE ADVANTAGE: POSITIONING *AND* CAPABILITIES

The concept of competitive advantage, as it has been widely used since the 1980s, has been proposed by Michael Porter[9] in his work on industrial competition. Porter's work survived the test of time not because he identified the definitive sources of competitive advantage, but because he provided a frame of reference, from which the debate could evolve effectively. Back in 1980 he identified two generic types of competitive advantage that companies can pursue: low costs and differentiation. Through the sheer efficiency of the use of labor and capital, a firm can obtain cost advantage by selling products or services with the lowest cost in its industry. On the other hand, the uniqueness of certain features of the products or services valued by consumers allows a firm to explore differentiation strategies. Among these unique features are the peculiarities of the product (e.g., its aesthetics, technology or performance), and the services provided by the firm, such as the technology employed in performing certain activities, and customer support. According to Porter,[10] to obtain competitive advantage companies need to have a clear strategy: "the creation of a unique and valuable position, involving a different set of activities".

A spontaneous question arising from Porter's generic principles of competition relates to their applicability to environmental issues. Can the pursuit of competitive advantage promote better environmental practices in firms? More specifically, what sustainability strategy best fit within the generic strategies defined by Porter? Addressing these questions requires one to recognize that the two traditional types of competitive advantage are fundamentally associated with the market performance of a company, which is ultimately translated into selling products and services.[11] In this perspective, the ability of a firm to trade high volumes of low-cost products or to obtain price premiums by selling differentiated goods or services represents a competitive advantage only if consumers value this capability in the marketplace. A simplified way of translating Porter's principles is to say that competitive advantage is a market advantage. In his perspective, "at the broadest level, success is a function of two areas: the attractiveness of the industry in which the firm competes and its relative position in that industry".[12]

The implementation and certification of Environmental Management Systems (EMS) by firms can serve as a test of Porter's perspective in the area of sustainability (Chapter 4 discusses the certification of EMS in more detail). Since the launch of the International Organization for Standardization (ISO) 14000 series in 1996 as a standard for EMS, academics and practitioners thought that the Total Quality Management (TQM) principles could be directly transplanted to the management of corporate environmental and social responsibility – from which the term Total Responsibility Management (TRM) emerged.[13] More importantly, they believed that the consequences would be similar. Based on this logic, the certification of a firm's EMS would not only be a systematic way of managing risks but would also have the potential to become a source of competitive advantage.[14] In the words of Sandra Waddock and Charles Bodwell: "responsibility management approaches can potentially provide for a solid basis of competitive advantage, especially for early movers".[15]

If competitive advantage is mainly a market advantage, why would the control of organizational processes via EMS become a source of competitive advantage? What kind of market advantage a certified EMS would create for the firm? After all, consumers buy products and services, not the activity system performed to deliver them, *per se*. While Porter's PS cannot provide a convincing answer, a rival school of thought in the area of strategic management can. For those familiar with the so-called RBV of the firm, competitive advantage is not a function of industrial structure but results from the ability of firms to use resources, which are heterogeneously distributed across competing firms and tend to be stable over time.[16] When compared to Porter's positioning perspective, the RBV does not constrain the choices available to firms regarding the structure of the industry; rather, it considers the competitive advantage resulting from the capabilities of firms to acquire and manage resources. Among them are technical capabilities, which includes those related to design for environment, ownership of intellectual property, brand leadership and financial capabilities, organizational structure and culture, all of which can be deployed to serve the goal creating competitive advantage around eco-innovation. In other words, instead of focusing on the external context, the RBV perspective highlights the influence internal organizational processes[17] exert on competitiveness. Ultimately, businesses create value by transforming raw materials into products via a multitude of activities, such as purchasing, transporting, manufacturing and delivery. Such influence explains the allusion made by academics and consultants since the early days of the *pays to be green* debate, that the certification of EMS would generate competitive advantage.

The specific case of environmental management in firms also highlights the growing importance intangibles have in the valuation of modern businesses. In fact, intangibles make clearer why the RBV and the PS should be seen as complementary perspectives, rather than rivals. The reason is simple: The way firms manage their activities have the potential to create or destroy value. For some eco-investments, organizational processes can have value in themselves (quasi)-independently of the position the firm has in the industry in which it operates, as well as from the value consumers attribute to its products and services. For instance, a firm with an ISO 14001 certification may have an advantage over a competitor simply because it satisfies the demand of a client organization – of having a certified EMS. Such advantage has little to do with the intrinsic features of the products sold by the company. In other words, the valuation of processes and products/services can occur independently.

Hence, specific circumstances favor companies to generate competitive advantage based on their competences in managing their organizational processes independently of the position that products/services have in the marketplace. Even so, because there is a close link between what firms produce (product/services) and how they produce (organizational processes), only by bringing the PS and RBV together one can identify and explain different sources of competitive advantage. Next section will use an analytical framework to simplify and further explore this proposition.

## COMPETITIVE ENVIRONMENTAL STRATEGIES[18]

Experienced consultants and academics have explored a multitude of examples to make the business case for eco-advantage and provide benchmarks for organizational greening.[19] Regrettably, most studies do not bracket situations in which such opportunities are minimal, implying that the potential to profit from eco-investments is equally distributed within as well as among firms, independently of the context in which they operate or their internal capabilities. Even if we accept that opportunities to profit from eco-investments are available to any business at different degrees, internal and external circumstances will facilitate or hinder firms to explore them. In other words, particular conditions favor firms to transform eco-investments into profitable business opportunities and, eventually, into sources of competitive advantage.

In order to identify such conditions, it is first necessary to decouple the elements involved in corporate environmentalism that have not been treated as independent areas of strategic action. This is exactly what the framework presented in Figure 2.1 does. The matrix combining these elements with the basic types of competitive advantage with which a firm seeks to achieve generates four possible strategies, represented in the figure. The quadrants represent a typology of specialized environmental strategies that corporations may adopt. The structure of

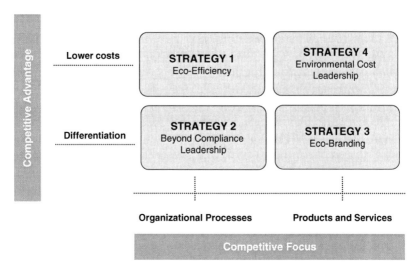

FIGURE 2.1 **Competitive Environmental Strategies**

the industry in which a firm operates, its position within that industry, the types of markets the company serves and its capabilities to acquire resource or to deploy radically innovative strategies will suggest the appropriate competitive focus (organizational processes or products/services) and the potential source of competitive advantage (cost or differentiation) for a firm.[20]

As it has been emphasized in the previous section, the distinction between organizational processes and products/services is possible only because of the fact that the four strategies can work independently. For instance, by engaging in a business code of environmental management or in a voluntary agreement[21] a firm may be trying to be a beyond compliance leader based on its processes, while its products or services may not present any clear environmental features, such as an eco-label. Conversely, a firm may decide to sell products with eco-labels but does not make explicit the green credentials of its activity system.

The framework has four subtle but very important facets. First, it should not be seen as a *stage model*.[22] Stage models normally focus on the societal needs for more eco-friendly corporate behavior, rather than on principles of strategic management to guide sustainability strategies, as discussed in this book. Stage models also presuppose that firms would go from low to high levels of ambitions. From the early 1990s when Charles Hunt and E. Auster[23] proposed a five-stage develop-ment continuum, ranging from beginners to pro-activists, till the more recent Sustainable Value Framework developed by Stuart L. Hart,[24] there is the presupposition that every company can go through increas-ingly higher greening stages. Even if these models work well on their own right, the one presented here does not assume an evolutionary pathway. The strategies depicted in the Competitive Environmental Strategies (CES) framework, do not present increasing degrees of com-plexity but rather are influenced and, therefore, applicable to certain conditions. In this respect, the CES framework is a *choice model*, rather than a stage model.

Second, in the CES framework, differentiation relates fundamentally to higher costs involved in managing organizational processes, as well as the willingness of customers to pay a price premium for products or services. However, as any stylization of reality, the strategies are ideal types of a particular phenomenon. After all, theoretical frameworks are means to help understanding reality by simplifying it. One should con-sider, for instance, that in every differentiation strategy (eco-oriented or otherwise), there is a cost component; and there is a differentiation component in every cost strategy. For instance, McDonald's and Burger King might compete on the basis of low costs, but by using renewable packaging material, for instance, McDonald's also tries to be different within the low-cost strategy. Differentiation might be pursued via eco-labeling but, as for many supermarkets in Scandinavia, there is price competition among eco-labeled products.

Third, there is an important difference between the scope of organi-zational processes (Strategies 1 and 2) and products/services (Strategies 3 and 4). Organizational and industrial processes are sets of inter-linked activities involving machines, equipment and people control-ling them, which are difficult to separate in practice. Products and services, on the other hand, are relatively easier to identify and iso-late (even assuming that services are activity systems). Because of this, organizational processes tend to have a more encompassing scope than products or services. For instance, it is possible to define a strategy for a single product or an eco-brand, such as the case of *Änglamark*, pre-sented in Chapter 5. While it is possible to do the same with a set of organizational processes, such as a specialized production line, more often they refer to the whole manufacturing facility or the even the entire corporation. In fact, the experience in developing the frame-work showed that, even though firms can have an all-encompassing sustainability strategy, more often they develop strategies for particular projects, product portfolios or strategic groups of products. The framework is particularly useful for such situations. It allows managers to think systematically about the various eco-investments they have to manage.

Fourth, it is fundamental to recognize that the focus on one specific strategy does not imply ignoring other areas in which environmental impacts can be reduced. For instance, if the company chooses to focus on Strategy 1 (Eco-efficiency), it does not mean that it should not try to reduce the impact of its products. After all, the framework departs from proactive, beyond compliance behavior (i.e., beyond what companies are required by law), and companies are expected to do their best to reduce their environmental footprint in as many fronts as possible. However, as in any other area of business, there will always be competition for resources. Developing projects A and B may imply that project C has to wait. Strategy implies choice, priority and focus. Among multiple possibilities, a chosen strategy implies that the company will do certain things (the certification of an EMS, for instance) and not others (an eco-label for its products).

Finally, a note on the title of the strategies: although we could identify the strategies only by their numbers, the title should help grasping their intrinsic logic. The drawback of titles is the meaning attributed to words and the risk of generating multiple interpretations. The title of Strategy 2, for instance could be applied to both organizational processes and product/services. After all, the entire framework presupposes beyond compliance behavior. But because beyond compliance behavior has emerged from environmental management in factories, more often it referred to the operations performed by industry. Overall, the title of each strategy should only facilitate the identification of their constituents but, in the same way the framework should not constrain thinking, the titles should not be taken literally.

## The Skeptical Corporate Environmentalist: Subtle Trade-Offs

As any tool, the framework will not be helpful if used in the wrong manner or for the wrong purpose. Once again, it is fundamental to recognize that the divisions between the four generic strategies represent a stylized typology to ease the identification of where the competitive focus of environmental strategies might be hidden. But rather than a straightjacket, the framework should be used as a frame or reference: the quadrants were proposed only to facilitate thinking and reduce the lack of clarity about environmental competitiveness. Because the distinction between competitive advantage based on processes and products/services is indeed a tricky one, it is quite easy to mix the general efforts companies make to reduce their environmental impacts (to be good) with the necessary focus a strategy needs to have in order to generate competitive advantages (to be the best). After all, there are conditions internal and external to the organization that will suggest the company to have a specific center of attention for its strategy. This is why the four quadrants are useful: they help us to think about the scope of proactive strategies and identify areas of priority according to the reality faced by the company. In other words, the division should help managers to make a strategic choice between different types of eco-investments, as well as to prioritize them in time.

Nonetheless, if boundaries between the four possible strategies are notional – a skeptical may ask – why should anyone consider the theoretical distinctions between processes and products/services? If products have to be produced in one way or another, and gains in process productivity can be transferred to products, would not this render the framework useless? Although subtle, there are indeed very practical reasons to make such distinctions. Finding opportunities beyond the "low hanging fruits" require a more detailed analysis of the elements involved in strategic environmental management. And, by definition, analysis refers to the breakdown of an issue into its components. In this case, the separation between processes and products/services allows one to identify the trade-offs between the strategic choices available to managers, mentioned previously. This is, according to Porter,[25] a fundamental condition for strategy.

Let us first analyze the distinction between Strategy 1 (Eco-efficiency) and Strategy 4 (E-cost leadership). Since the optimization of an industrial process can contribute to the reduction of the final price of a product and increase its chance to compete in the marketplace, one could claim that any distinction between processes and products does not make any sense. This would be correct if eco-efficiencies in processes were enough to make a product an E-cost leader. Unfortunately, this is rarely the case. E-cost leadership requires a firm to achieve both: the lowest cost and the lowest environmental impact of the products in its category. As Chapters 3 and 6 explore in more detail, substantial gains in process-oriented resource productivity are only part of the equation; it has to be matched with eco-design, product dematerialization, substitution or new models of commercialization. In many firms, the reduction of both the final cost of the product and its environmental impact is a result of product redesign and substitution of raw materials, rather than Eco-efficiency of processes. In other words, competitiveness can be obtained by focusing on the nature of the product, rather than on the process.

A firm may have competences to champion eco-efficiencies during production and eventually cut the final cost of its products. These products may still, however, present high levels of environmental impacts and, therefore, would not be able to become E-cost leaders. In fact, this is the case of the vast majority of (mass) volume producers today. Economies of scale in manufacturing have forced companies to optimize their systems of production and use similar processes for a wide range of products. In the car industry – with hardly any good example of eco-friendly vehicles in the market – platform sharing became a common practice in the 1990s. Today, gains in process productivity will impact on a wide range of car models, not just one.

Considering the ecological attributes of most eco-oriented products rely on their mode of production, some may also raise an objection for the distinction between Strategies 2 and 3. Undeniably, a selected number of companies have been able to link the environmental qualities of their products with organizational processes, resulting in corporate-wide eco-differentiation. From the outset, pioneering companies such as the Belgium Ecover, a manufacturer of cleaning products, the British Body Shop and the Americans Ben & Jerry and Patagonia have been able to establish a close link between the ecological responsibility of organizational processes and the portfolio of products sold by them. For these firms, green credentials were foundational values, which are so pervasive in whatever these companies do that greening became an intrinsic attribute of their products. The history of these companies suggests that it is indeed possible to adopt corporate-wide eco-differentiation, thus blurring the distinction between Strategies 2 and 3. They also suggest, however, that such strategic scope is restricted to a small number of firms born with eco-oriented values and commitments. Although a late conversion is possible for some firms, linking product brand to corporate environmentalism might prove a difficult trajectory for most. And, even when this is possible, the distinction between processes and product/services remains a useful analytical approach. The eco-investments of General Electric (GE), the giant American producer of a wide range of products, and the Japanese carmaker Toyota make this point clearer.

Toyota *Prius*, the successful Hybrid Vehicle (HV) launched in 1997 (discussed in detail in Chapter 7) and GE's *ecomagination* program launched in the early 2000s had a clear focus on the products of these companies. The eco-investments received so much media attention that both companies ended up being recognized as sustainability leaders in their industries. As Chapter 5 discusses, it is natural that some strong brand image confound products with the overall company, and for both GE and Toyota, linking a successful ecological initiative with the corporate image was certainly welcome. But while the ecomagination program generated institutional eco-differentiation, the value of GE's brand, which has historically been among the highest in the world,[26] can neither be attributed to its late conversion to corporate environmentalism, nor to the success of its eco-oriented products. In the case of Toyota, sales of hybrid cars are is still marginal when compared with the more polluting models based on internal combustion engines, rendering it difficult for Toyota to justify its green image. On the other hand, since the 1980s, Toyota has been so efficient in its production processes that the term *Toyotism* emerged almost as synonymous of TQM. Once again, the distinction between organizational processes and products/services is useful to bring up these kinds of intricacies involved in the analysis of sustainability strategies.

The distinction is particularly useful for the analysis of other subtleties between Strategies 2 and 3. Many of the ecological attributes of eco-labeled products, for instance, are located in their methods of production – more specifically, in the agricultural (i.e., organic food) or industrial (i.e., unbleached paper for coffee filters) production processes. When considering the purchase of a pack of wheat flour, the consumer is evaluating the intrinsic features of the flour, which has been influenced by its mode of production. For some skeptics, the relationship between the characteristics of the products and the processes used in their production may suggest that the distinction between Eco-branding (Strategy 3) and Beyond Compliance Leadership (Strategy 2) is irrelevant. A close look into the example of the wheat flour proves otherwise. When buying the flour at a supermarket, the consumer is not judging the organizational processes of the supermarket but the producer of the flour. For consumers, it would make little difference if supermarkets have the ISO 14001 certification of its processes, for instance. From this perspective, the distinction between the organizational processes and products is crucial for the optimization of eco-investments made by the supermarket (in this case, have a portfolio of eco-labeled products).

Finally, managers or academics may also question the distinction between cost and differentiation of organizational processes (Strategies 1 and 2, respectively). They may argue that the work toward eco-efficiency can facilitate the development and reduce the implementation costs of an EMS. Conversely, a firm might use its ISO 14001 certification to differentiate from its competitors, but the process of implementing the EMS might uncover potential areas in which an increase in resource productivity can be obtained.[27] If the strategic choices reinforce each other, what would be the reason to establish a distinction between competitive advantage based on cost (Strategy 1) or differentiation (Strategy 2) of organizational processes? None, if opportunities for double dividends or win–win scenarios were always prevalent. But the reality is that EMS certification is often costly and opportunities for cost savings are exceptions rather than the norms. Confronted with the trade-offs between the relatively higher costs associated with EMS certification (differentiation), and the focus on eco-efficiency of industrial processes (costs), managers will have to choose between Strategy 1 and 2. If the conditions for process-oriented differentiation are not available to the firm, the choice of focusing on Eco-efficiency would emerge naturally from the analysis based on the framework.

## BEYOND COMPETITION: SUSTAINABLE VALUE INNOVATION

Some large organizations exert powerful influences over their supply chain and the distribution channels of their products and services. Companies such as Wall Mart, the giant American retailer, can request via policies and management standards not only what upstream suppliers have to do in order to gain contracts, but also how products have to be produced, and at what prices they should be sold. Including social and environmental standards into the selection criteria of suppliers is certainly easier for Wall Mart than for an average medium-size retailer. This is what has been termed *the greening of supply chain*. Firms can also influence downstream practices. The case of Tetra Pak discussed in Chapter 1 is a typical example of excellent *product stewardship*. By investing in activities that eventually facilitate a close-loop recycling system, Tetra Pak reduces the risk of being taken by surprise, in case a take-back regulation is imposed. Nonetheless, in that case, even though such practices may affect reputation positively, because they are non-rival, competitive advantages are not expected to emerge from them. The entry of a competitor in the recycling network established by Tetra Pak will represent a gain for everyone: more recycled material (public benefits) and, since expenses would now be shared, lower cost (private profits), to maintain the network.

Such upstream and downstream influences can be important components of the Corporate Environmental Strategies (CES) of companies like Wall Mart or Tetra Pak. After all, green supply chains and product stewardship can enhance the position of these companies in existing industries – retailing and packaging, respectively. Chapters 2 to 6 explore the issues involved in the definition of CES, which assist companies to compete in existing industries and markets. To use the metaphor of the proponents of the Blue Ocean Strategy, CES help managers "swim in red oceans". The four CES presented in Figure 2.1 are well suited to a vast number of companies that have little room to exit existing industries. In essence, these strategies follow the logic of rivalry adopted by Michael Porter's PS and the RBV of the firm. The main concern of corporations adopting CES relates to the creation of competitive advantages, so as to increase their market share in well-established industrial sectors. By adopting such strategies, suppliers of products and services within a specific industry try to outperform each other so as to have the biggest possible "share of the cake". These are, essentially, *supply-oriented strategies*.

Highly innovative organizations, however, can bypass competition altogether via Blue Ocean Strategies (BOS). By redefining the value proposition for customers, firms can eventually create value innovation – differentiated products and services at low prices.[28] In other words, BOS eliminates the trade-off between cost and differentiation, present in competitive (environmental) strategies. Since the development of BOS is based on the customers needs, rather than on what competitors do to outperform each other, it can be considered a *demand-oriented strategy*. In other words, value innovation is created by satisfying untapped customer demands with new value propositions. Since new market spaces do not compete with existing offers, there are no comparative standards for pricing, blurring the traditional distinctions between low price and differentiation of supply-oriented strategies.

The central question arising from BOS relates to its applicability to ecological sustainability in business. Can the demands of customers be aligned with societal demands for environmental responsibility? How value innovation can be created in an ecologically sustainable manner?

How sustainability strategies can contribute to the creation of new market spaces? In order to answer these questions, it is first necessary to recognize that value innovation resulting in higher environmental impacts does not restrict the deployment of BOS. On the other hand, sustainability strategies require embedding the value proposition into the demands of other stakeholders for environmental protection and social justice. Translating BOS into sustainability results in a strategy that simultaneously lower costs, increase consumer value and generate public benefits. Such requirements lead us to the concept of Sustainable Value Innovation (SVI), depicted in Figure 2.2.

By questioning the appropriateness of the business model in creating extensive value – not only for shareholders but also for society – SVI redefines the boundaries of the value system of an existing industry. By presenting a value proposition that is so unique – normally via a new business model – companies can reduce both economic costs and environmental impacts, while generating value not only for customers but also for the society as a whole. Chapter 7 employs the auto industry to explore the intricacies of this highly demanding sustainability strategy, and to explain why SVI is, in essence, a systems strategy, for it requires changes in both the nature and technology of products and in the logic by which systems of production and consumption are organized. Obviously, such type of strategy is by far

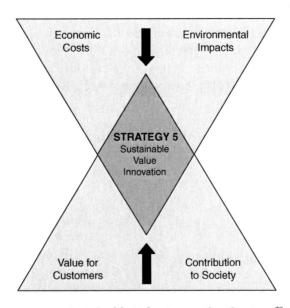

FIGURE 2.2 **Sustainable Value Innovation Strategy**[29]

the most difficult to be deployed. Nonetheless, companies that are able to develop a radically innovative approach to the management of the life cycle activity system of its products, can create value to existing and new customers at both reduced economic and environmental costs. By doing so, SVI strategies align shareholder pressure for profits with societal and environmental demands.

## THE SUSTAINABILITY STRATEGIES PORTFOLIO

Although the logic embedded in Competitive Environmental Strategies (CES) (Figure 2.1) and Sustainable Value Innovation (SVI) (Figure 2.2) is fundamentally different, some similarities require a brief explanatory note. E-cost Leadership (Strategy 4, presented in Chapter 6), is relatively close to the principle of value innovation, since products/services are leaders not only in price but also in their environmental prerogatives. In the case of E-cost, however, consumers refuse to pay premiums for eco-differentiation, so whatever eco-attributes products may present, what defines the sales of the product is its low price for the client; the company is constrained to focus on low costs, exclusively. Low pricing, in this case, is fundamental for the firm to gain competitive advantage in existing markets.

In the case of Eco-branding (Strategy 3, presented in Chapter 5), differentiation is central for the marketing and sales of the products/services but, in that case, it refers mainly to the willingness of customers to pay a price premium for them. Differentiated value is created at higher costs, so the company has no choice but to focus on existing niche markets that are willing to pay relatively higher prices. Hence, in order to be competitive in existing industries, CES require a clear choice between low costs (E-cost Leadership) and price premium (Eco-branding). Such constrain is not present in the logic of the SVI strategy. Since value innovation creates new markets, price comparisons are blurred, leaving SVI strategies with more scope to explore middle-ground prices. In this case, "to be stuck in the middle", rather than being a problem, it is actually an advantage.

Together, CES and SVI compose the five possible sustainability strategies. Figure 2.3 schematically presents these strategic choices, emphasizing the broad conditions for deployment. Companies can use CES, described in detail in Part II of the book, to enhance the competitive positioning in existing markets, while SVI strategy is deployed for the creation of new market spaces (Part III). In Chapters 3 to 7, the

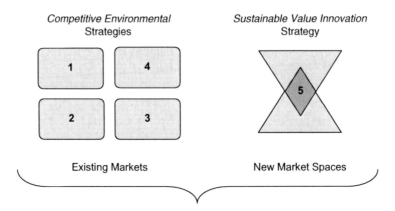

FIGURE 2.3 **Sustainability Strategies**

answers for *when it pays to be green* are summarized in two dimensions: the external context of organizations, such as the industry in which they operate, as well as their internal capabilities to deploy any of the sustainability strategies.

Finally, this chapter is intended to make clear that sustainability strategies are choices available to managers to align environmental and social investments with the generic strategy of the company. By focusing on a specific strategy, managers choose to prioritize certain eco-investments in detriment of others, which may be deployed in a later stage. However, the focus on Eco-branding (Strategy 3) does not imply that the company should make no efforts to increase the efficiency of its processes, for instance. After all, sustainability strategies depart from behavior that goes beyond what is required by law. Companies adopting any of the five strategies are already doing well in the environmental and social fronts. What the choice does imply is an alignment of eco-investments with the context and competences of the company, as well as with its generic strategy. By using clear criteria to focus on a particular sustainability strategy, managers are in a better position to justify eco-investments to shareholders, while addressing the demands of other stakeholders.

## CONCLUSION

This chapter applied the fundamentals of strategic management to identify what sustainability strategies are and are not. Although much has been written on the topic, a closer scrutiny shows that what are commonly argued to be strategies are, often, generic approaches to environmental management in firms. The ground rules of Michael Porter's views on strategy and the overall recognition that strategy involves trade-offs and choice, was used to substantiate this point consequently. A closer look into the uniqueness of environmental issues in business also indicated the need to establish a distinction between competitive advantage based on organizational processes and products/services. Although subtle, such distinction is fundamental not only for the identification of unambiguous Competitive Environmental Strategies (CES), but also approximates the antagonist Positioning School (PS) and the Resource-Based View (RBV) perspectives. In addition, the concept of value innovation by Kim and Maubourgne, was extended to the realms of sustainability management. The fifth sustainability strategy – Sustainable Value Innovation (SVI) – laid the basis for the creation of new market spaces that are aligned with the demands for environmental and social responsibility.

The categorization of five sustainability strategies constitutes a solid step toward the identification of conditions in which eco-investments can increase the competitiveness of firms or create new market spaces – broadly encompassed by the question: *when does it pays to be green?* By decoupling the main constituents of the debate around this question, the chapter initiates the journey toward a better understanding of environmental issues in business, as well as the design of unambiguous strategies. Such theoretical refinement, anchored in solid empirical research, is timely. Managers who have increasingly been requested to respond to demands to ecological sustainability need to base eco-investments on solid grounds. They have been bombarded with theoretical approaches, tools, techniques, schemes, standards and demonstration cases of best-environmental practices. While the multiplication of material helps to bring environmental issues to mainstream business thinking, a great number of managers are wondering what they should do first, or why. Should they focus on generating carbon credits via eco-efficiencies? Enhance corporate reputation by joining Green Clubs? Subscribe to eco-label programs? Explore emerging cleantech markets? Develop an eco-brand? Move from selling products to less-impacting services? Within the confusion, many do a bit of everything, and spend precious resources without sound rationales.

After the emergence of sustainability as a management topic, what managers need is a basis from which they can prioritize eco-investments. More broadly, they need to align these investments with the generic strategy of the company. This is why the research resulting in the frameworks, concepts and ideas presented in this book is useful. Academics may use the frameworks presented in this chapter and explored in detail throughout the book to identify theoretical constructs or empirical variables for the definition of research propositions or hypotheses. On more practical grounds, the classification schemes can help managers to define and prioritize areas of organizational action, optimizing the overall economic return on eco-investments, as well as the potential to transform these investments into sources of competitive advantage and new market spaces.

# PART II
## COMPETITIVE ENVIRONMENTALISM

# 3

## ECO-EFFICIENCY

At different degrees, all organizations pursue resource productivity. By optimizing the overall use of resources, such as the reduction of energy consumption and waste, companies can also reduce the costs associated with them and, consequently, become more competitive. However, as emphasized in Chapter 2, even if resource productivity is conducive to higher competitiveness, it cannot be considered a strategy *per se*. Resource productivity is part of the overall operational effectiveness that companies need to perform in order to remain competitive. If this is the case, why is then Eco-efficiency portrayed as a generic type of competitive environmental strategy in p. 30? If resource efficiency is a well-known management practice that every organization should pursue, would it not be just operational effectiveness? Or is there something fundamentally new about the ecology-oriented efficiency?

Yes, there is. By focusing on three new aspects of resource productivity, this chapter explains why this is the case. The first relates to the consequences of adopting Lean Thinking (LT) in factories. Eco-efficiency strategy can lead to breakthrough innovations and radical improvements in resource utilization. By applying LT to operations management, wastes and by-products can eventually be converted into new sources of revenues. Extended beyond the borders of the organization, such logic can yield further gains via Industrial Symbiosis (IS). Finally – and more importantly – lower emissions of carbon dioxide ($CO_2$) via eco-efficient strategies are becoming increasingly lucrative, affecting the competitiveness of corporations in an unprecedented way. Solutions to global warming encompass new market mechanisms that will progressively reward eco-efficient strategies.

## FROM RESOURCE PRODUCTIVITY TO ECO-EFFICIENCY

In the early 1990s, Porter[1] re-emphasized that productivity is the key element for companies to gain competitiveness. Organizations should be able to transform costs into profits by identifying concealed opportunities for innovation, leading to more efficient organizational systems. In later work, Porter and Clas Van der Linde[2] asserted that companies should promote resource productivity in the form of materials savings, increases in process yields, and better utilization of by-products because waste consists, fundamentally, of an inefficient use of resources. Companies would only need to find hidden opportunities to profit from eco-investments and eventually transform them into sources of competitive advantage, they argue. As mentioned in Chapter 1, the *win–win hypothesis* triggered an intense debate during the 1990s, which is by far still open, as this book suggests.

The publication of the *Natural Capitalism* in end of the 1990s represented another milestone in the debate about resource productivity, competitive advantage and sustainability. In their book, Paul Hawken, Amory and Hunter Lovins[3] demonstrated that by redesigning systems of production and consumption, the potential gains on resource productivity are so considerable that a new economic system may emerge. They substantiate their argument by presenting examples of corporations that are increasing the productivity of natural resources, shifting to biologically inspired production models, moving to a solutions-based business model, and reinvesting in natural capital. Such practices would promote natural capitalism – what capitalism might become when regulatory and market mechanisms eventually succeed in making organizations to internalize environmental costs.

Although the *Natural Capitalism* caused some stir in business and political circles in America and Europe, environmental issues would soon be overshadowed by the dominance of social aspects of business, as mentioned in the Preface of this book. The anti-Kyoto Protocol standing of the George W. Bush presidency somehow transformed environmental problems into second-class citizens. In both academic and business circles, the environment was put under the umbrella of Corporate Social Responsibility (CSR). While corporations started issuing CSR reports, scholars and consultants shifted their attention from environmental to social issues in business. The turning point of the social-centric trend came after a sudden awareness about the dangers of climate change. Global warming finally spoke to the general public via dreadful hurricanes and the tireless personal crusade of Mr Al Gore, which rendered him the 2007 Nobel Peace Prize. Whatever caused this tipping point, it brought back the urgent need for societies to reduce the anthropogenic impacts on the natural environment. Suddenly, the atmosphere became more receptive to corporate environmentalism. While global warming became a common language among politicians and even lay citizens, more down to earth issues such as eco-efficiency returned to the jargon of scholars, consultants and executives.

Eco-efficiency has been, in fact, part of the corporate vocabulary for quite a while. In particular, the dissemination of the eco-efficiency terminology among business people is due to the (over) simplified concept proposed by the World Business Council for Sustainable Development (WBCSD) in 1996: "doing more with less". The definition is certainly catchy for business but it does not reflect the true scope of eco-efficiency. More troublesome; it equals the concepts of eco-efficiency and resource productivity.[4] Although resource productivity is a chief way of lowering the environmental impact of industrial processes and products, it is only part of it. For instance, since the 1970s automobile emissions of toxic substances (known as $NO_x$) were lowered by 97 percent. Such performance was obtained by improving engine technology, suggesting that eco-efficiency is far broader than resource productivity.

Besides process optimization, eco-efficiency can also be obtained by substituting or reducing the amount of material used in a product – hence the *dematerialization* jargon. More broadly, ecological efficiency can be achieved by reducing the environmental impact of the whole system of production and consumption, as well as all the stages prior to manufacturing and post-consumption.[5] Hence, an improved generic definition of eco-efficiency would be: "doing more with less and lower environmental impact". Following this logic, a more precise title for Strategy 1 would be *Process-Oriented* Eco-efficiency. But since the concept has more often been associated with process improvements than with product innovation, Eco-efficiency seems sufficiently adequate (and less clumsy) to describe it. In the framework, Eco-efficiency Strategy encompasses the notion of resource productivity throughout the life cycle of products, suggesting that the search for process-oriented efficiencies is based on ecological prerogatives.

By focusing on Eco-efficiency strategies, firms aim at reducing both the economic costs and environmental impacts of organizational processes. Since cost reduction is crucial, most companies working on Strategy 1 do so without much fanfare. For instance, with the aim of reducing their environmental impacts and risks, SMEs (Small and

Medium Enterprises) may develop an Environmental Management System (EMS), but because they simply do not have resources to spare, these companies may avoid the costs associated with EMS certification. Implementing a much simpler and less-bureaucratic EMS than the ones using the guidelines of ISO 14001, may simply make business sense. Firms supplying a relatively small number of industrial markets (i.e., supplying other firms, also known as business-to-business – B2B) may avoid the costs of EMS certification by simply inviting their client organizations to informally audit their systems.

By their very nature, most process-intensive firms are expected to pursue the Eco-efficiency strategy. Less obvious, however, is the possibility of generating new business opportunities out of what has formerly been considered waste. Overall, firms focusing on Strategy 1 are expected to develop capabilities to continuously increase the productivity of organizational processes while decreasing both the environmental impacts and the costs associated with them. However, in order to go beyond operational effectiveness – hence, to become a real strategy – companies have to not only present the lowest operational costs in its industry but also obtain extra value for wastes, by-products and resource productivity itself. The next sections identify three main ways of achieving such aims.

## ECO-EFFICIENCY AT THE FIRM: *LEAN THINKING*

Resource productivity in manufacturing has always made business sense. In the context of automobile assembling, for instance, business survival has demanded constant efforts to reduce the costs of industrial processes. Rationalization of systems of production became an imperative for car manufacturers just to remain competitive. The pressure to cut costs in every possible manner has driven auto assemblers to work toward the minimization of waste and optimization of resource at its maximum. Overall, the reduction of the environmental impact of car manufacturing can be seen as a consequence of the Lean Thinking (LT) approach prescribed by James Womack and Daniel Jones,[6] in which companies should eliminate any source of inefficiency – particularly in the form of waste and by-products. Departing from this logic, eco-efficiency epitomizes LT and the search for radical-resource productivity.

If this is the case, would not Eco-efficiency become operational effectiveness, as described in Chapter 2? Undeniably, carmakers should not expect to obtain competitive advantage out of eco-efficient strategies.

The auto industry is well known for its high levels of process optimization, and zero-waste practices are ubiquitous in the industry. However, automakers should not expect to generate competitive advantage out of Strategy 1 simply because eco-efficiencies have become the average operational effectiveness of the industry (or their license to operate). On the other hand, in industries with low levels of resource efficiency, first movers can indeed obtain competitive advantages – at least for certain time. As the average performance of the industry increases, however, eco-efficient strategies tend to converge to operational effectiveness.

Once again, does this mean that Eco-efficiency is always prone to become operational effectiveness? If this was correct, Eco-efficiency simply could not be considered a strategy. The reason why this is not the case lies on the spin-off effects of an eco-oriented mindset. As Amory Lovins and associates have asserted so sharply, Eco-efficiency strategies can lead to breakthrough innovations in technology, allowing quantum improvements in resource utilization. LT can trigger eco-innovations with the potential to become sources of competitive advantage. In practice, when wastes are not considered as such in process-intensive firms, extra revenues are made possible by taking full advantage of by-product utilization. Although the development of such opportunities require considerable management capabilities, creativity, and as in any business activity are doomed to failure, they are readily available to some business.

According to the Zero Emissions Research Initiative (ZERI),[7] such potential is clear in breweries. Since the mid-1990s, eco-efficient practices in beer production has been a leading case of the network. In a traditional beer factory, spent grains, which represent the larger part of the breweries' by-products (circa 18 kg per hectolitres of beer), are sold to pig or cattle farmers at low prices. Although this makes economic sense for the breweries, it is neither ideal for the animals nor environmentally sound. Because the cattle cannot properly digest the spent grain, they produce excessive methane in the process – a gas contributor to global warming. As an alternative to such practice, because grains are rich in fibers and protein, ZERI proposes them to be used to farm mushrooms. With relatively unsophisticated equipment, it is possible to separate the enzymes generated by the breakdown of lingocellulose and the protein-enriched substrate resulting from the process of mushroom growing. Besides being a more environmentally friendly solution, mushrooms have higher market value than animal fodder. The resulting five categories of enzymes can also be used as additives in soaps.[8] Overall, by recovering the protein, which has traditionally been considered waste in the beer industry, breweries could generate new sources of income and, eventually, new businesses.[9]

If such practices are possible, some skeptics could ask why only a few small breweries in developing countries have adopted them? Why growing mushrooms out of spend grains has not become normal practice in the industry? Part of the answer relates to what corporations consider as their core competence.[10] Beer makers are rarely interested in peripheral activities, such as waste management. Even less if such activities demand new capabilities, investments and, when compared with much larger revenues coming from their core business, have relatively low returns. For a large beer producer such as Heineken, whenever possible, spent grains are sold as cattle feed. For Heineken, this is a satisfactory solution from both economical and ecological points of view. But besides being sub-optimal as cattle fed,[11] this solution is not applicable when cattle farming is too faraway.[12] In the current logic of breweries, even less applicable is the idea of growing mushrooms. Location and scale seem to matter. Technically, producing mushrooms out of beer's by-products might make business sense for a micro-brewery located in areas where there is demand for animal fodder, and where any opportunity to create (unqualified) jobs and extract additional value – even if marginal – of a business is always welcome. Such firms normally supply local markets and are more prone to be operating in developing countries. As for most multinationals of the beverage industry, rather than waste from grains, water has increasingly become the main environmental concern.

The Chinese Guitang Group, an industrial conglomerate that operates one of the largest sugar refineries in the country, presents a more compelling example of generating (extra) value out of resource optimization. By installing downstream companies to utilize nearly all by-products of sugar production, the group not only reduced environmental emissions and disposal costs but also improved sugar quality and generated new revenues. The Guitang Group consists of interlinked production of sugar, alcohol, cement, compound fertilizer and paper, including recycling and reuse of by-products and waste. "By choosing to approach its waste as a business opportunity, the Guitang Group solved a traditional problem by using the sludge as the calcium carbonate ($CaCO_3$ – most commonly known as chalk) feedstock to a new cement plant. This, in turn, generated profits that helped offset the higher cost of the carbonation and increased the company's competitiveness in the sugar market."[13] Such performance entails enduring management challenges. In order to maintain its competitiveness in the global sugar market, while optimizing the system, the Group has to influence continuously both downstream and upstream operations. It guarantees its supply base, for instance, through technological and economic incentives to farmers. Overall, for more than four decades the Guitang Group has been implementing an Eco-efficiency strategy that resulted in an inter-organizational network with special features. The industrial system supports some of the fundamental concepts of industrial symbiosis.

## ECO-EFFICIENCY BEYOND BORDERS: INDUSTRIAL SYMBIOSIS

As the Guitang Group example hints, the eco-efficiency mindset can generate gains beyond the borders of a single firm. Companies can reduce environmental impacts and costs by looking for resource efficiency of broader processes. From the perspective of Industrial Ecology (IE),[14] individual manufacturing processes are constituents of broader systems of production and consumption. In such systems, the total cycle of materials – from virgin material, to component, to product, to waste product and to ultimate disposal – is optimized in terms of resources, energy, environmental impact and capital.

Although the scope of IE encompasses the entire life cycle of products – hence, from pre-manufacturing to post-consumption – the metaphor has inspired the development of Industrial Symbiosis (IS) more strongly than any other area. IS presumes an interdependent flux of materials, processes and energy inside the cluster where firms are located. In practice, waste, by-products and energy from one firm can feed processes in another, forming closed-loop systems, commonly called Eco-Industrial Parks (EIPs). In such parks, firms collaborate for an eco-efficient management of materials, processes and energy inside the cluster. The economic logic for such practice is simple: either the costs of inputs and wastes can be reduced or income can be obtained by selling by-products or energy to neighboring firms.[15]

The EIP in the town of Kalundborg[16] emerged in 1970s and was first documented as an eco-park in 1989. Even so, it remains the benchmark for EIPs. In this Danish city, the need to assure the long-term supply of fresh water triggered a process that would become the point of reference of IS. A coal-fired power plant, an oil refinery, a pharmaceutical company specializing in biotechnology, a gypsum board and plastic board plant, concrete producers, a producer of sulfuric acid, the municipal heating authority, a fish farm, some greenhouses, local farms and other enterprises, all cooperate in order to optimize the use of energy and resources as well as to reduce the waste. The basic idea behind the system is that wastes and by-products of one company can become raw material for another. Without using any government regulations, these organizations established contracts for an efficient flow of materials and energy. Similarly to the Guitang Group, as the companies optimized the use of resources and reduced their environmental impact, their internal costs decreased. But there is a basic difference between the two cases: while the Chinese IS was formed under the umbrella of one single-owner organization, the participant companies in the Danish EIP belong to various owners.

One of the reasons for the Kalundborg EIP to be known worldwide relates to the possibility of lowering production costs by adopting novel ways of dealing with waste and the optimization of inputs and outputs of materials and energy. This is an appealing argument for business and a reason for governments to promote collaborative projects mimicking the Danish case. In the United States (US) for instance, during the Clinton administration in the 1990s, the US President's Council for Sustainable Development (USPCSD) promoted 15 EIPs. An equal number of parks were assisted by other entities in the US, while over 30 were classified as eco-parks in Europe – mainly in the Netherlands, England and Sweden.[17]

By 2007, the results were not so encouraging. An evaluation of the eco-parks promoted by the USPCSD showed poor results. Out of the 15 cases, only one was considered a standard eco-park; the remaining projects never emerged, opened with a different goal, or simply failed. There were idiosyncratic reasons for such failure; many related to the lack of funding or changes in the political priorities and preferences of local governments. Nonetheless, such failure triggers a more obvious question: considering the alleged benefits of eco-parks, why have they not succeeded in such a business-prone context such as the US? Part of the answer has its roots in the regulatory framework. Regulations tie firms to their waste indefinitely. Whenever companies manufacture products out of waste from another firm (secondary material) this firm could be held liable for health problems or damages caused by the waste used as raw material. Hence, the obvious attitude is to avoid getting involved with other company's waste.

More broadly, the experience of promoting eco-parks in the past two decades uncovered fundamental differences between industrial and natural systems.[18] As we can intuitively guess, industrial clusters do not function as organic entities, as ecological niches do. Operational, financial, behavioral and political issues limit the willingness of companies to participate in IS schemes. Low price of resources and waste disposal constitute a major economic barrier, and the costs of waste disposal tend to be buried in the overheads. Some companies also do not want to run the risk of exposing information they consider commercially sensitive or confidential. And, once again, IS runs against a business trend initiated in early 1990s. Business schools have been teaching MBAs and executives not to deviate from the core competences of their companies. According to this logic, unless you are in the waste management sector, dealing with waste is simply not your business.

Does this mean that the IS was born dead? After all, if the business case is still weak, why should anyone bother? Managers should bother because the main limitation of EIPs relates to public policies, rather than private management. What became clear in the assessment of eco-parks is the limitation of (top-down) public programs to trigger collaborative networks.[19] Since the essence of collaboration is rooted in trust, a formal government program rarely generates the results of a natural alliance (bottom-up), as it was the case in Kalundborg. In other words, business enterprise is what matters most in such schemes. Rather than relying on government intervention, managers with an eco-efficiency mindset can look for solutions wherever solutions can be found; even if they are beyond the borders of the company. From the business perspective, it is not important whether synergies are conducive to an ideal EIP. This is rather an academic concern. Instead, what matters for them is the possibility for synergies to push eco-efficiencies beyond the average performance of the industry.

The case of Kwinana in Western Australia is an exemplary in this respect. In this cluster of primary mineral industries, 32 by-product and 15 utility synergies were identified.[20] They were a result of factors that are unique to the local context. Essentially, because local firms mainly supply international markets there are low levels of inter-firm competition. The location also matters. The cluster is isolated from other industrial centers but relatively close to the metropolitan center of Perth in Western Australia, which is rapidly encroaching around the area. The proximity of the industrial district to a sensitive marine environment and recreational area for residents is also promoting IS. There has been increasing public awareness and valuation of the natural resources in that area. Finally, the Kwinana Industries Council has played an important role in addressing issues common to the companies and promoting collaboration among firms.

The Kwinana case leads to a counter-intuitive conclusion. Many would think – as most academics and government officials did – that it would be easier to develop EIPs at the drawing board stage. In trying to replicate the Kalundborg EIP, it was thought that symbiotic synergies would be more difficult to happen in areas in which the infrastructure is already heavily in place. But the Australian minerals industry cluster suggests the opposite. Once the companies are running relatively well, focusing on their core competences, they seem to be more willing to further explore resource productivity and eco-efficiencies in a broader, regional scale.

There is no reason, however, to ignore the potential of planning eco-industrial systems. As long as it is feasible to minimize opposition, an EIP can certainly emerge out of a well-conceived business plan.[21] The least risk-prone way of developing it is to put all organizations under the same ownership, as cooperatives in the agro industry do. But while there is nothing stopping agro industrialists to make investments similar to the Chinese Guitang Group, such privately owned symbiotic systems are expensive to install and difficult to coordinate. Under a cooperative organization, the system can be more flexible and adaptable to market changes.

Overall, for process intensive firms, it makes business sense to consider IS as a fundamental component of their Eco-efficiency strategy. Similarly, companies dealing with agricultural produce have good reasons to pay special attention to synergic practices. As the production of bio-fuels is expected to grow substantially in tandem with crescent need for food crops, the integration of farming and industrial processes is imperative for reducing both the environmental impacts and economic costs. The Brazilian Pro-alcohol Program is exemplary in this respect (discussed in more detail in Chapter 6). From 1980 to 2005, the costs of ethanol production fell from US$100 to US$ 30 per barrel due to improvements in agricultural techniques and, more significantly due to the use of sugarcane bagasse (a by-product of sugarcane crushing) for energy production, avoiding the use of any fossil fuel in the industrial project.[22] Besides, agro-IS has the potential to yield higher profits and higher environmental and social performance, as the Guitang Group project suggests. More broadly, IS can promote the economical and technical feasibility of new and arguably more sustainable production and consumption systems.[23] But the potential does not end in the synergies. As the next section poses, such projects can obtain profits from an increasingly important source.

## ECO-EFFICIENCY IN THE SKIES: CARBON CREDITS

The distinction between operational effectiveness and eco-efficiency is subtle and, for many, it might indeed look like just good house-keeping. But the reason this is not the case relates to the potential eco-efficiencies have to generate more than traditional cost savings. Taken to their extreme potential, LT and IS can lead to gains that go beyond reducing the costs via operational effectiveness and, there-fore, become a strategy *per se*. Nonetheless, if such extra dividends still do not seem so obvious, they become crystal clear when we con-sider the potential of eco-efficiency to generate a radically new type of value; one that goes further than the physical resources managed by companies. Mitigation measures for global warming – the most cru-cial environmental issue of the twenty-first century – will increasingly reward Eco-efficiency in a novel way. For some companies operating in energy-intensive industries, Strategy 1 may generate extra value via carbon credits.

Machines and their embedded technologies have always been the means of producing other things, such as equipment or consumable products. For instance, the price of a set of machines for packaging depends on its productivity, normally measured in number or packs per standard period of time. Waste generated during calibration or operation enters the processing cost account, so machines can be com-pared on the basis of their cost-efficiency ratio. Similarly, the costs of installing filters with the aim of reducing the emissions of pollutants have served either to comply with legislation or to demonstrate the willingness of corporations to reduce their environmental impact. But until recently, very seldom companies could mitigate such costs.

The market mechanisms embedded into the Kyoto Protocol par-tially changed this situation.[24] The process goes back to 1988, when the United Nations (UN) established a group called Intergovernmental Panel on Climate Change (IPCC). In its first report in 1990, the IPCC emphasized the seriousness of climate change threat. The vast major-ity of the two thousand scientists involved in the panel agreed that the release of Greenhouse Gasses (GHG)[25] by human activity is warming the planet at an unprecedented scale. The first IPCC report was timely for the Earth Summit that would be held in Rio de Janeiro, Brazil, in

1992. The report was instrumental to catalyze the international community to consider collective action to combat climate change. As a result, 166 nations, (which grew to 193 in 2008),[26] signed the United Nations Framework Convention on Climate Change (UNFCCC), with the not legally binding target to stabilize their emissions of GHG at the 1990s levels by the year 2000, which came into force in 1994.

Only when the second IPCC assessment report was released in 1995, calling for strong policy action, that political action was mobilized toward what eventually became the Kyoto Protocol.[27] Mandatory reduction targets would be set at the follow up UNFCCC conference, in December 1997 in Kyoto, Japan. The conference in Kyoto became a milestone and one of the most controversial events in the early history of mitigation policies for climate change. After round-the-clock negotiations and the assistance of the US Vice-president Al Gore, the parties agreed to reduce the levels of emissions by 5 percent, compared to the 1990 levels, in the commitment period 2008–2012. Developing countries were exempted. The targets are different for different countries, with some such as Norway, Australia and Iceland having even the right to increase emissions. For the treaty to become law, it was necessary that at least 55 percent of the Annex I nations ratified it. Hence, even though the US never did and Australia ratified it only in 2007, the protocol became international law in 2005, after Russia's ratification. Because there were great uncertainties whether or not Kyoto would become law, most signatory governments did not take significant measures to be prepared for the commitment period, and reaching the targets became a major challenge for most; Canada leading the league.

The market mechanisms woven into the Protocol – international Emission Trading Schemes (ETS), Clean Development Mechanism (CDM) and Joint Implementation (JI) projects – value the potential of certain technologies reduce emissions of GHG.[28] The central argument for implementing these mechanisms relies on lowest-cost compliance under the Kyoto Protocol. When it comes to international emission trading, although the European Union (EU) has been the world leader, several American states developed carbon registration schemes or passed or proposed legislation, independently of the official position of the federal government, which has opposed the Kyoto treaty. Japan also developed a voluntary ETS, resulting in some demand for credits generated in developing countries. Canada and Australia opened their voluntary ETS in July 2007. Although there are some differences, because they are all under the Kyoto protocol, most ETS work in a similar way. The EU-ETS serves as an example. The scheme is based on the *cap and trade* principle in which the *cap* is a fixed total amount of emissions that EU member countries can release. This, in turn, defines how much groups of companies and individual firms operating in five energy-intensive industries can emit. In the first period (2005–2008), allowances were given for free to companies working in five sectors: electricity, oil, metals, building materials and paper.[29] Such firms could trade any carbon credits (which are, more precisely, non-emitted carbons). On the other hand, firms that exceeded their emissions, in order to remain under the quota, had to buy emission credits in the market.[30]

Under the Kyoto protocol, JI projects are carried out in industrialized countries, while developing countries host CDMs.[31] Typical eligible projects are the generation of renewable energy, fuel switching and energy efficiency. Such projects represent opportunities for investors interested in generating carbon credits via eco-efficiencies or alternative energy sources. They may find more opportunities in CDMs because the geographical scope is larger than JIs. As an indication, in 2008 there were only 175 JI projects, while close to 4,000 CDMs[32] were registered. Hence, for local firms or subsidiaries of multinationals operating in emerging markets, CDMs represent an opportunity to generate extra dividends via Eco-efficiency strategies.[33] For instance, some of the eco-efficiency gains by the Guitang Group, mentioned in the previous section, also qualify for CDMs. Overall, the possibilities for CDMs are surely vast.

The encapsulation of landfills with subsequent sequestration and burning of methane to generate electricity has been a typical CDM project but methane can have other origins. For instance, Bunge, a giant American agricultural commodity corporation and Sadia, one of the largest producers of poultry, pork and beef in Brazil installed biodigesters in Brazilian pig farms. Sludge ponds were encapsulated for the collection and subsequent methane burning to generate electricity for the farm. Because the resulting $CO_2$ is relatively less damaging than the methane, such projects are entitled to carbon credits. Sadia and Bunge finance the investments to the farmers and retain the majority of the credits, which are then sold to firms operating in Annex I countries. In 2006, Sadia sold the credits obtained in four of such projects to the European Carbon Fund (Luxemburg) for €35 million. Finally, methane can also generate carbon credits via IS. For instance, Shell, the Anglo-Dutch oil producer, is pumping $CO_2$ from a refinery in the Botlek area in the Netherlands into 500 greenhouses producing fruit and vegetables, thus avoiding the emissions of 170,000 tons of $CO_2$

per year and saving the greenhouse owners to burn 95 cubic meters of gas to produce the $CO_2$ they need.[34]

Two additional cases in Brazil illustrate how CDMs can be extracted from eco-efficient industrial processes.[35] SA Paulista,[36] a Brazilian company operating in environmental management and urban infrastructure, created a system to use methane released in a landfill to generate electricity for the city of Nova Iguaçú, in the state of Rio de Janeiro. In November 2005 the credits were sold to the Dutch government for €9 million. The Brazilian subsidiary of Rodhia, a Swiss multinational, installed filters in the Paulinia factory in the State of Sao Paulo, so to reduce the emissions of Nitrous Oxide ($N_2O$) in the production of nylon. In March 2006 Rodhia sold its carbon credits (more precisely, in this case, $N_2O$) to the French Société Générale and IXIS Bank for €54 million.

Although certification and trading of project-based instruments under CDMs and JIs encompass a good deal of bureaucracy, the logic does not differ from other commodity markets. In the words of Andrew Hoffman:[37] "the entrepreneurial question in GHG reductions is: How can one generate carbon credits at the lowest cost and sell them at the highest price?" Firms operating in the EU can sell any excess of emission reductions via the EU-ETS while, those located in developing countries as the projects in Brazil suggested (i.e., not assigned emission reductions), can sell emission reductions to EU countries via CDMs. Overall, under market-oriented regulatory frameworks, corporations that develop competences to innovate and reduce emissions at relatively low costs can certainly improve their competitiveness.[38]

Carbon markets in Europe and CDMs in Asia and Latin America are well established but emerging carbon markets face multiple challenges. There are many uncertainties about calculation methodologies, institutional capacities, such as organizational competences for monitoring the development of projects and registration bodies, as well as transaction and abatement costs. Despite these uncertainties, the trend toward political commitment to GHG reduction indicates that market mechanisms will increasingly reward carbon reductions beyond the Kyoto commitment period. In simple terms, competences to reduce emission levels of operations turn out to be a strategy *per se*.[39] Overall, there has been a clear trend toward market-oriented incentives for the de-carbonization of the world. And even if there is no clarity after 2012, when the first commitment period of the Kyoto protocol expires, the uncertainties are less about whether there will be a new emissions plan and more about how it will look like. For business, whether the post-Kyoto regime will be a global treaty or an array of decentralized alliances is not as important as whether carbon trading will prevail as a market mechanism. Indeed, the trend points toward the consolidation of carbon credits as one of the most important commodities of the twenty-first century.

## WHEN *ECO-EFFICIENCY* PAYS

If opportunities for profiting from eco-efficiency are so readily available, one could question why they are not common practice in most business? – Mainly because managers lack the time and capabilities to focus on the flow of resources within and around the corporation.[40] Eco-efficiency also requires specific conditions to be developed, and therefore cannot be extended indistinctively to all firms. Although at some level gains can be achieved in virtually every organization, particular circumstances will reward some more than others. Similarly, for each Competitive Environmental Strategy (CES) the answer for *when it pays to be green* depends on the context in which firms operate, as well as in its internal competences.

### Context

The organizational context can refer to several aspects. The industry in which the company operates is the most obvious, but it also includes the political and cultural settings embedding the firm. In this respect, companies based in stable industrial democracies face very different demands than those for operating in countries in which social unrest is common. The types of markets and clients served by the firm are also important contextual dimensions. Overall, specific external conditions favor or hinder companies to draw positive outcomes from eco-investments.

Generically, empirical evidence suggests that Eco-efficiency strategies have greater potential to generate competitive advantage in firms supplying industrial markets (B2B), facing relatively high levels of processing costs and generation of wastes and/or by-products. Many firms in the agribusiness and the food and beverage industries fall into this category. In such circumstances, because the client organization is not willing to bare extra costs associated with environmental protection, focusing on Eco-efficiency strategies makes business sense. By working toward eco-efficiency within the firm, as well as beyond its borders, process-intensive firms will be saving money while decreasing the environmental impact of their operations.

Curiously, the electronics sector is one in which the potential to profit from eco-efficiencies is significant. Although the sector has often been identified as high-tech, chip-making factories have been designed so poorly that 100-plus percent improvements are possible in most cases.[41] Manufacturing microchips has a great potential to benefit from reducing the use of water and the emissions of Perfluorinated Compounds (PECs), a GHG with high climate impact.[42] There is also evidence that Eco-efficiency strategies are of particular importance for the minerals and chemical industry, as the Kwinana case suggested.

Regarding Industrial Symbiosis (IS), because organic matter is less prone to be toxic, the potential to transform by-products and waste into inputs for new industrial process is greater in the agribusiness than in most industrial activities. In fact, symbiotic relationships have been part of agriculture since its early days. Synergies between pastoral agriculture and the production of cereals and animal protein have historically been a common practice among (mainly small) farmers worldwide. Such practices only changed with the advent of the modern industrial farming, in which monocultures are grown with the help of petrochemical fertilizers and pesticides.[43] But monocultures based on chemical inputs are increasingly questionable from the point of view of both risk avoidance and resource optimization. Monoculture of corn and soybeans, for instance, has only increased the dependency of farmers to the fluctuations of commodity prices, and has caused waves of misfortunes in Brazilian farms since the 1970s. As the production of bio-fuels to power motor vehicles is pushed forward at a global scale, a balanced integration of crops and industrial production seem to be more sustainable from both economic and environmental perspectives.

Eco-efficiency strategies represent a special opportunity for energy-intensive industries via several de-carbonization incentives. Besides the mechanisms woven into the Kyoto Protocol, governments have designed additional incentives for corporations to take a proactive stance regarding energy and climate change issues. Market mechanisms were put into place for specific sectors, such as the Tradable Green Certificates (TGC), created in Europe for the energy sector.[44] In general terms, the creation of a TGC scheme entails a mandatory renewable electricity target that energy producers are required to meet during a given time period. The underlying purpose is to encourage a cost-effective deployment of renewable energy technologies on the energy supply side. Green certificates are given to companies that present evidence that they produced energy (electricity in particular) using renewable sources, such as hydro, wind, geothermal or solar. They can trade the certificates to meet their individual targets and equalize marginal compliance costs. Companies with a surplus of green certificates can benefit from the additional profits obtained from the sales. Buyers of green certificates benefit from the cost savings, as the price of purchased certificates is lower than the marginal costs of producing additional in-the-house renewable energy.

On the other hand, Tradable White Certificate (TWC) schemes have been created to increase energy efficiency on the energy demand side.[45] The creation of a TWC scheme entails a mandatory energy savings target that certain market actors (usually energy suppliers) are required to meet during a given time period. Like in any tradable certificate scheme, flexibility is critical because it allows companies to choose how to meet their targets cost effectively.[46] France, Great Britain and Italy have established the market to trade certificates resulting from energy efficiency measurers, such as the replacement of appliances and lighting for energy-efficient ones, as well as insulation improvements in households and commercial buildings. The realized energy savings are credited with certificates. The key market strategy for actors depends on the market price of TWC compared to the costs of realizing their own energy savings. Firms with a surplus of certificates can save them for future commitment periods or to speculate on rising market prices.[47] Although TGC and TWC schemes are not ubiquitous, the trend signals their adoption (or at least interest) by most Annex I countries, and by some developing countries. Since environmental credentials became important for the general public, governments are doing their best to implement market-based policy incentives.

## Competences

In order to explore Eco-efficiency strategies, companies will need to develop competences for Lean Thinking (LT), Industrial Symbiosis (IS) and carbon trading. As long as there is genuine willingness to gain skills, there is nothing stopping companies to learn them. But because developing competences might take some time and effort, the sooner the work starts the better. As described in the work of Womack and Jones, developing LT is not as straightforward as some imagine, and only a few corporations in the world have actually managed to become so lean, as prescribed by the authors. After all, the whole idea behind LT is to compete only against perfection: "Our earnest advice to lean firms today is simple: To hell with your competitors; compete against perfection by identifying all activities that are *muda* (waste in Japanese) and eliminating them. This is an absolute rather than a relative standard, which can provide the essential North Star for any organization". According to the proponents of natural capitalism:[48] "the message and method are stark: don't study it, just do it, keep trying. If you fixed it, fix it again". Things are not that different for IS. As it was highlighted in the previous sections, successful Eco-Industrial Parks (EIPs) have little to do with planning and lot with trial and error.

Companies will need to develop competences not only to generate carbon credits but also to make the most in trading them. In the first allowance period of the European Emission Trading Scheme (2005–2008), for instance, many manufacturing companies did not take full advantage of carbon markets because they view the ETS as a regulatory burden, rather than a chance to make money. Because environmental experts were put in charge of trading, the main focus was on making sure the company had enough allowances, rather than maximizing value, as financial whizzes would do.[49] By contrast, energy utilities operating in deregulated markets are prone to have developed such skills while trading power and adjusting their power sources based on the changing costs of different fuels on a daily basis. These firms are certainly in a better position to explore Eco-efficiency strategies. Experienced in-house traders provide the competences needed to variable emissions-price adaptation with relative ease; competences that utilities operating in regulated markets will have to develop. This is also true for trading in the TWC market. In order to fully benefit from the commercial opportunities that this policy provides, there is the need for companies to incrementally build competences in these new market mechanisms.[50]

Overall in order to develop Eco-efficiency strategies, managers need to either have answers to the following set of questions, or use the questions to guide them through potential projects within the organization. Although the list can obviously be much longer, managers can start with these ones:

- How eco-efficient are our manufacturing processes? How do we measure it? Do we have rewarding mechanisms to stir eco-innovation in our company?

- Do we have any LT program? Do our people understand the principles of resource productivity and eco-efficiency?
- Does the company generate much waste and by-products? Can we utilize these as an input for other processes? How much? Can waste be reused, remanufactured or recycled internally or by other firm? If so, to what extent are we doing that? Can we improve it significantly?
- Do we have competences to explore IS with neighboring firms? Do they have similar problems with waste management? Are they willing to exchange information with us? Do they have anything we could use in our company? How can we build trust with them?
- Do we have the competences to create an internal ETS so as to motivate people to work toward eco-efficiency, as well to learn more about its possible mechanisms?
- What is our potential to generate carbon credits from your operations? Do we understand the ETS and their intricacies? Is anyone in charge?
- Does anyone in the company have competences to operate in an ETS? Overall, what is our strategy for carbon?

## CONCLUSION

For quite some time, the potential to profit from resource productivity has been emphasized by experts. Time and again they pointed out the enormous opportunity companies have to increase resource productivity and eco-efficiency. The widespread use of cases in the literature suggests that there is indeed a place for the double dividends in most areas of economic activity. What experts have seldom done, however, is to bracket the specific contexts in which such opportunities are less prone to occur. Although some firms can indeed achieve Factor Four improvements (i.e., with 75 percent less resources, energy and environmental impact), others are more limited by the nature of the business. In other words, even though some levels of improvements are always possible, they are not evenly distributed across firms and industries. What has been lacking for quite some time is the mapping of such opportunities.

This chapter contributed to such mapping by explaining why focusing on cost reductions via Eco-efficiency simply makes business sense for process-oriented companies (mostly operating in industrial markets). Such focus, however, does not imply that the company should not make any efforts to reduce and communicate the overall environmental impacts or processes, products and services. It means only that, for some firms, the possibilities of generating competitive advantage out of process-oriented Eco-efficiency are higher than in other sustainability strategies. Although companies should do their best to address stakeholders' expectations, such efforts should not detract them from their strategic focus. By focusing on the Eco-efficiency strategy, they decide not to spend excessive energy in areas in which competitive advantage is not prone to emerge. As the next chapter will explore in detail, building reputation out of investments for GHG reductions, for instance, is very unlikely. Because returns of eco-investments depend on the context in which the firm operates and its capabilities, managers have to decide what the company should do first, and why.

Sustainability strategies should be seen in the context of relational choices. Once managers recognize that their business activity privileges Eco-efficiency strategies, they can concentrate their efforts toward the exploration of hidden opportunities. As this chapter has shown, the rewards will come in the form of lower operational costs and extra revenues, such as the transformation of by-products and waste into new business and the generation of carbon credits. Such economic advantages constitute the fundamental rationale for the deployment of Eco-efficiency strategies – even if risk reduction and improved corporate citizenship can surface as additional outcomes. In fact, for some companies facing strong public pressure, reputation is so important that risk management and stakeholder dialogue becomes the main rationale of their sustainability strategies. For such companies, being good is simply not enough. They need to look good in the eyes of the stakeholders.

# 4

## BEYOND COMPLIANCE LEADERSHIP

Some companies operating in resource-intensive industries are often in the radar of eco-activists. For such firms, it is vital that customers and the general public know about their efforts to go beyond compliance. They spend money to certify Environmental Management Systems (EMS) according to International Organization for Standardization (ISO) 14001 or subscribe to other voluntary environmental initiatives, which may require them to pay membership fees and make commitments to reduce the overall impact of organizational processes. Companies in the energy and oil business, for instance, are among those that have been spending millions not only to go beyond compliance but also to advertise their green credentials. These companies have been doing their best to demonstrate their commitment to reducing the impact of operations, going beyond what is required by law so that demanding stakeholders see them as leaders in environmental protection.

The rationale behind such efforts is simple: reputational risk. Corporations exposed to risks of bad reputation need to demonstrate their credentials of good citizenship. Some learned the hard way. They faced major reputational crisis because of industrial accidents and consequent pressure from eco-activists. Large resource-intensive corporations had to invest heavily to move from risk avoidance to building positive reputation, which could eventually shelter them from stakeholder pressure. Besides making large eco-investments for safer industrial processes, they also invested substantial resources to reposition themselves and learn about stakeholder dialogue and engagement. For this purpose, they adopted a new form of regulatory practice, which involved a great degree of voluntarism and engagement with civil society organizations. Often, they also anchored their claims of corporate environmentalism in codes and standards of environmental management.

Back in the 1990s, many believed that practices such as the endorsement of codes of environmental conduct or the certification of ISO 14001 would generate competitive advantage. Some firms indeed profited from early certification. But since then, the number of Voluntary Environmental Initiatives (VEI) or Green Clubs – as they are also known – mushroomed, populating a murky area of a wide variety of voluntary initiatives. Today we reached the stage in which it is necessary to ask: Why and when should a company join a Green Club? When does membership help businesses to avoid reputational risks? Do Clubs facilitate stakeholder dialogue so that firms can build positive reputation over time? On more practical grounds we can ask, for instance, when does an ISO 14001 certification make business sense? What type of industries and market segments are more prone to reward EMS certification? Overall, what are the necessary conditions for firms to differentiate on the basis of organizational processes? When does this kind of effort pay off?

This chapter addresses these questions by inquiring into the essentials of reputation and the role Green Clubs have in shaping it. Although much has been said about how companies create reputational value, this remains very controversial; even more so when we link reputational value with corporate environmental and social responsibilities. An exemplary case of stakeholder dialogue opens the discussion for the identification of what is, ultimately, corporate environmental reputation. The chapter then reviews the main initiatives promoting beyond compliance practices in corporations, developed in the past two decades. Hence, besides the underlying logic of using Green Clubs as instruments to build Beyond Compliance Leadership strategies, the chapter presents historical developments that can help readers to have a better overview of such initiatives.

## REPUTATION AND ITS RISKS

Southern Pacific Petroleum (SPP) was founded in 1968 with the main purpose of discovering and exploring oil shale deposits in Australia.[1] Exploring oil in the form of black tar-like substance called bitumen, which sticks between certain sands and shale, has been technically feasible for quite some time. But extracting and processing oil shale has historically been more expensive than conventional (liquid) oil. For SPP, the long-term development of the deposits in Queensland had the potential to meet the needs, and even create an export market for

Australia. The oil shale industry could provide a bridge to meet modern energy needs until a cleaner source of fuel is developed, which is viable and affordable. SPP made the following statement:[2]

Oil provides 92 percent of the fuel required for our cars and trucks and is the essential fuel for our planes and boats. For at least the next 15 years, Australia, along with all other developed countries, will be highly dependent on oil for all major transportation needs. Whilst governments around the world are investing heavily in developing new fuels that are less greenhouse intensive, this is expected to take at least a generation. In the meantime, we all remain dependent on oil.

The need for oil and the proven Australian oil shale reserves justified the design of pilot plant in 1990 with government support in the form of an excise rebate. In 1995, Suncor Energy, a Canadian firm secured €159 million (A$250 million) for a joint venture with SPP to build an oil shale demonstration plant in Australia, known as the *Stuart Oil Shale Project*. Suncor Energy has been a pioneer in the exploration of oil sands in Canada, and it is currently the single largest producer. The company is also involved in natural gas exploration and production, as well as refining and marketing operations.

The first stage of development in the Stuart Project had the capacity to produce 4,500 barrels of oil per day. After the second and third stages, the venture had the objective of increasing production to some 85,000 barrels of oil per day. Although the pilot plant was completed by 1999, long before it was targeted by eco-activists. The location of the project near the Great Barrier Reef, a very sensitive marine environment, the significant greenhouse gas emissions associated with oil shale, and the potential impacts of global warming on coral reefs and human activities attracted the attention of local stakeholders and Greenpeace. Exploration, production and transportation operations put pressure on the natural environment and local communities, not to mention the global impacts of fossil fuels on climate.

Not surprisingly, Greenpeace initiated action against the project immediately. By highlighting the long-term impact of the project on the natural environment, as well as championing local issues such as community health, the activist organization was instrumental in rallying intense local support behind their cause to see the project terminated. There were indeed many submissions from local stakeholders rejecting the continuation of the project: More than one hundred local residents took legal action for €7.4 million (A$12 million) in compensation for health problems and reductions in property values. Seventy of these local residents claim to be affected from the dioxin air emissions. Around 14,000 people and 17 environment, tourism and fishing groups lodged written submissions to the State Government opposing the proposed expansion of the plant, a record for public submissions to the State Government. A total of 27 environment, tourism and fishing groups jointly submitted an assessment to the Fuel Tax Inquiry, calling for the Federal and State Governments not to approve any further development or increased financial support. Finally, the Environmental Protection Agency (EPA) stated in 2001 that the Stuart Oil Shale Project was a public health nuisance.

As a result of eco-activism pressure, in 2001 Suncor Energy had no other choice than withdrawing from the Stuart Oil Shale Project. The official statement from the company was the need to focus on projects in its home nation of Canada. However, press releases from Greenpeace purported that it was their campaign to educate Suncor Energy and the intense public reaction about the unsustainable industry, rather, that led to the retreat.

In May 2002, the Chairman of SPP, Campbell Anderson, pledged their commitment to improve environmental performance in both their current and future plants, and submitted several initiatives asking for SPP to be judged on their ability to resolve their environmental problems, once identified. SPP argued that the Stuart Oil Shale Project had achieved a great deal in being able to improve their operations to ensure that they met or were below relevant environmental guidelines for emissions. But an investigation carried out in 2003 by regional development specialists, indicated that the environmental impact had been underestimated, and that there was the intractable problem of a noxious industry adjacent to a horticultural community. The pressure from Greenpeace had been forceful from the outset, combined with the public opposition forced the Stuart Oil Shale Project to be terminated in 2004.

Curiously, the closure of the project coincided with the prices of oil that makes the exploration of oil shale economically competitive.[3] But the bad stakeholder dialogue dented the chances of Suncor to explore the vast Australian reserves. Moreover, the debate over the Stuart Project brought the operations of oil sands in Canada to the attention of the general public, which could eventually become the next target of eco-activism.

## Making a difference: the *agency* lesson

*It takes many deeds to build a good reputation, and only one bad one to loose it. (Benjamin Franklin)*

The Stuart oil shale case in Australia is an extreme example of bad stakeholder dialogue, which was opportunely explored by Greenpeace. Although the project had an undeniable local environmental impact, the precautionary measures surrounding the project were of high standards from the outset. In fact, if we consider the full life cycle of the product explored by the SPP, by far the largest impact was not local but global. It happens via the burning of the oil (from shale) consequently contributing to global warming. Nonetheless, the arguments brought against the project partners referred mainly to the impact that the local operations (organizational processes) would have in the local community. Hence, it is fundamental realizing that the project failed not because of its bad environmental credentials *per se*, but because of the bad perception the community had about them. Greenpeace, being experienced on such types of campaign, explored the issue wisely and transformed a local concern into a national crusade. At once, it was a relatively easy win from Greenpeace.

The Stuart oil shale case uncovers the negative perceptions many of us have about the oil and mining industries. Historically, corporations have exposed local communities to all sorts of dangers to their health, local sustenance base and security. Some mining projects became icons of social unrest and environmental degradation. For example, in the 1990s the Ok Tedi, a mining company whose 52 percent of the shares were owned by the Australian giant BHP, faced legal charges for polluting the Fly River beyond recognition in Western Papua New Guinea. The local community, who saw their livelihood from fishing in the river disappear, won the dispute and got A\$110 million in compensation. The Ok Tedi company ended up paying the compensation, but it tried to influence the local government to change the legislation, which would make illegal for local villagers to sue mining companies following the Fly River incident.[4] Although BHP was not the single shareholder and therefore not the only offender, the incident badly damaged its reputation.

Such examples of disrespect for local population of humans and other species made people afraid of what comes out of mines and chimneys of factories, and the risks of having firms processing minerals, oil and gas in their backyard. Most industry experts and executives have been aware of such negative externalities for quite some time. But real action was taken only when some accidents that made history threatened the license to operate of some Multinational Corporations (MNC).

The Australian oil shale case and the OK Tedi mine in New Guinea put forward another lesson: when organizational processes are brought to the attention of consumers and general public, in the vast majority of the cases this influence is negative. It depends on what sociologists call *agency*: anything that provokes changes in the perceived reality of people and mobilizes public opinion.[5] In simple terms, agency is the ability to make a difference. Greenpeace was instrumental in influencing the public opinion about the Stuart project. Australians are very proud of the copious biodiversity of the Great Barrier Reef and the idea of a large dirty project killing the small and colorful fish of the Reef caught their imagination. Greenpeace transformed itself into an agency. Such reaction is similar to the well-documented case of the fight between Greenpeace and Shell in the North Sea. In that occasion, European consumers rejected the decision of Shell to dump oil rigs in the North Sea by boycotting its products.[6] In that particular circumstance, consumers' disapproval of Shell's practices was felt through a sharp slump in sales.

In fact, boycotting has historically been one of the few effective powers held by consumers to force business to act more responsibly. This kind of behavior has been classified as negative ethical consumerism,[7] since consumers screen companies and then boycott their products for the negative views they have about these firms. As the North Sea case suggests, boycotts tend to have a spectacular effect that can dent the corporate reputation quite badly. Even if measuring such effect is not straightforward, it is quite easy to understand why the net effect is negative. For this reason, many large companies operating in extractive industries realized they should do their best to avoid becoming targets of eco-activists and run the risk of having their reputations damaged. And, as the Stuart oil shale case shows, the overall environmental performance of companies operating in pollution-prone industries is becoming increasingly important to stakeholders. Hence, an explanation for the radical change of Shell's communication strategy is simply reputational-risk avoidance. Shell joined a group of companies that is trying to go beyond "avoiding the negatives" and instead working toward more stakeholder dialogue and engagement. Before addressing such initiatives, however, it is necessary to have a clear understanding about corporate reputation and how it can be measured.

## Corporate reputation: the perception of responsibility

*Character is like a tree and reputation like its shadow. The shadow is what we think of it; the tree is the real thing (Abraham Lincoln)*

So much has been written about reputation, yet it remains a controversial topic. Reputation is one of those concepts that people immediately have an idea and an opinion about, but as soon as we try to converse we notice different meanings. This is because, in Wittgenstein's[8] terms, the meanings of reputation have a family resemblance to each other. Similarly to the way in which members of a family may resemble each other but with no single core element that can tie them together, the use of the concept of reputation tends to mean different things to different people – including management academics and practitioners. Hence, a practical way of looking at reputation is to understand the perceptions of a corporation's responsibilities in the eyes of the stakeholders, as Hillebrand and Money creatively did in their research.[9] They analyzed, for instance, how a particular company relates to *me* as a customer through the kinds of benefits it offers via products and services; how it communicates with *me* and if it keeps promises. How the company relates to *other* stakeholders, such as the local community by being aware of their problems (some might even be caused by the company) and how, if at all, it acts to solve these problems. Finally, reputation is also about how the company relates to *itself* (shareholders) in terms of financial performance and long-term business success.

Besides this proposition, there are several well-established indexes to measure corporate reputation. They differ on the overall underlying approach, the stakeholders surveyed and the criteria companies are judged on. Nonetheless, all major indexes demonstrate a strong proximity between corporate reputation and corporate responsibility. For example, the Fortune's Most Admired Companies (MAC) list evaluates reputation according to eight aspects that are most admired by journalists, financial analysts and CEOs. Innovation, financial soundness, employee talent, use of corporate assets, long-term investment value, social responsibility, quality of management and quality of services and products are all criteria for responsible management – economic and social. Another example is the Reputation Quotient (RQ) index, which describes reputation as a result of stakeholder expectations. The RQ is based on six pillars of reputation: emotional appeal, products and services, vision and leadership, workplace environment, financial performance and social responsibility. Other indexes worth mentioning are the Corporate Personality Index, which surveys customers and employees about the *personality* of a company, and the SPIRIT index, which is the most comprehensive in terms of number of stakeholders surveyed and the dimensions of the index.[10]

Overall, corporate reputation relates to the perception stakeholders have about the firm. If we assume a black and white viewpoint, bad behavior would lead to bad reputation, motivating companies to do their best to be good and to satisfy stakeholders' expectations. But reality is not black and white. Often, there are competing demands form different stakeholders, and it is not clear-cut identifying the most powerful or the amplitude and legitimacy of their claims. Managing reputation is extremely complex, which explains the often-bombastic examples of well-managed firms falling from grace. As we saw in the previous session, the political stances of some civil society organizations can badly damage corporate reputation, independently of what the real facts are; after all, reputation is about perceptions, not reality.

Reputation depends largely on a third opinion or on the influence of opinion setters. Good corporate reputation can hardly be built on what the company says about itself. Besides trying to be responsible in all possible spheres of action, there is so much a firm can do alone to influence its image. Legitimacy is highly dependant on third party endorsement and verification. This explains why so many companies have joined Voluntary Environmental Initiatives' (VEI) or Green Clubs, as some prefer to call them (henceforth used interchangeably), such as the Coalition for Environmental Responsible Economics (CERES), Responsible Care, Global Compact, Climate Leaders, or standards for EMS, described in the next sessions.[11] To use a generic academic term, these initiatives are special types of decentralized institutions "because participation is voluntary and because diffused actors, rather than a central authority, provide rewards for participation or sanctions for not participating"[12]. In principle, they are instruments used by companies to better communicate with stakeholders, often serving as a shield against bad reputation. Companies have been spending significant resources to engage in Green Clubs and legitimize their beyond compliance practices, and eventually differentiate from competitors.

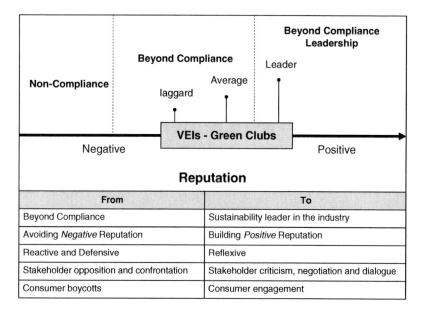

FIGURE 4.1 **Corporate Environmental Reputation and Leadership**

Figure 4.1 depicts the rationale for these companies to use Green Clubs as means of enhancing stakeholder dialogue, communication and engagement, which will eventually affect their reputation.

The figure also shows the main logic of the remaining sections of this chapter. Basically, companies have used Green Clubs as instruments for the management of their environmental reputation. They tried to move from the left to the right borders of the figure – from avoiding negative to building positive reputation; from stakeholder confrontation to dialogue; and from consumer opposition and boycotting to engagement. Overall, the following sections explain how companies have learned to adopt a reflexive position toward sustainability issues in business; rather than a defensive stance, characteristic of the last two decades or so. Many learned by becoming targets of eco-activism, but this should not diminish the importance of their late achievements. Their reframing and more recent attitude also contributed to corporate environmentalism to become a more widespread phenomenon. Moreover, by inducing firms to reduce their environmental impact, some Green Clubs generated both public benefits and private profits in the form of better corporate reputation. The origins and intricacies of the most important initiatives help to clarify why this is the case.

## *GREEN CLUBS*: REPUTATION INSURANCE?

In the 1980s, a series of disasters influenced public opinion to demand industries to approach environmental issues from a different perspective. A new mode of addressing regulation started to emerge, and a new phase of industry-governments relationship gradually gained terrain.[13] Green Clubs originally emerged as a means of helping firms to manage their reputation, which was often damaged by accidents or local pollution caused by their operations. Distinct sectors of the business community and non-profit organizations responded to the public demands for better corporate performance with the release of a series of voluntary initiatives, in the form of codes of conduct, environmental guidelines, charters and programs.[14] These initiatives share the common objectives of assisting business to guide the implementation of environmental programs, and to communicate this commitment to the general public. At the same time, such voluntarism detracted governments from imposing more restrictive regulations because club members would accept private costs to generate (non-excludable) environmental benefits. Green Clubs:[15]

> Require club members to incur private costs, as codified in the club's membership standards and mechanisms for ensuring compliance with those standards. The costs of membership should not be trivial because producing a public good is not free. For Green Clubs, the main costs of joining the club are generally not direct payments to the club sponsors. Rather, they are the monetary and non-monetary costs of adopting and adhering to the club's requirements.

Since the 1980s there has been an explosion of Green Clubs, even if only a few are relatively known by the general public. The ones briefly described here and depicted in Figure 4.2 are among those few, which are both historically relevant and exemplary in determining whether they generate any advantages for the member companies.[16]

The Canadian Chemical Producers Association (CCPA) proposed one of the first sector-specific initiatives that gained international reputation. The Responsible Care was created in 1983 as a voluntary program for companies in the chemical industry. In December 1984 public outrage at the disaster in the pesticide production plant of the Union Carbide in Bophal, India,[17] induced CCPA to make Responsible Care a condition for association membership. But it was not until 1988 that the American Chemical Manufacturers Association (CMA)

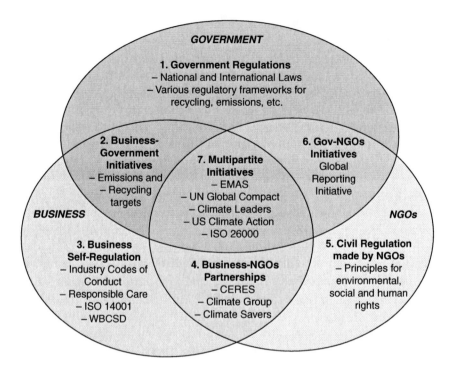

FIGURE 4.2 **Multi-stakeholder Regulatory Initiatives**

adopted the program, which was later also embraced by the European Chemical Industry Council and other chemical associations around the world. Responsible Care, which is basically a voluntary initiative for health, safety and environmental performance improvement, can also be seen as a self-regulation by industry, specifically designed for the chemical industry. In other words, differently from traditional regulatory measures imposed by governments (Item 1 in Figure 4.2), or emission targets agreed by industry and government (Item 2 in Figure 4.2), Responsible Care is self-imposed regulation by the members of the industry (Item 3 in Figure 4.2). Business self-regulation is also the ethical framework around which member companies operate, representing their commitment to respond to public concerns about the safe management of chemicals.

The extension in which Responsible Care is implemented varies significantly from country to country. In Canada and Australia, the United States, Brazil, South Africa, Austria and New Zealand third party verification has been instituted, but this is not the common practice for most countries - specially the ones where specific codes of environmental practice have not yet been developed. Apparently, the codes can become a useful instrument for the improvement of relationships between chemical companies and their main stakeholders. Even if the environmental performance improvements have been uneven among members, Responsible Care has enhanced the reputation of companies operating in the chemical sector.[18]

There is a multitude of comparable sector-specific initiatives, such as the International Council on Mining and Metals,[19] which had its (implicit) original rationale to protect the industry against bad reputation. Overall, we can expect that the vast majority of sector-specific VEI will result in a license to operate, rather than in competitive advantages for the participating firms. As the Responsible Care showed, as a whole, the reputation of members have been mostly protected or even improved. However, because the program is extensive to all players within the industry, no manufacturer obtained advantages over its rivals.

If by being a member of an industry-specific VEI companies avoid competitive disadvantages do they gain advantages in joining a VEI that is not sector-specific? The case of CERES is illustrative in this respect. The Exxon Valdez oil vessel accident in Alaska in 1989 prompted environmental groups and American institutional investors to launch the Valdez Principles, subsequently renamed the CERES Principles.[20] Such initiative characterizes what became to be known as *civil regulation*,[21] or the regulations that emerges from civil society organizations (Item 4 in Figure 4.2). In the early stages, proactive companies such as The Body Shop and Ben & Jerry, which already had strong environmental reputation, adopted the Principles, but in 1993 Sun Oil became the first Fortune 500 listed company to join in. By 2008, several large organizations, including General Motors (GM), Ford, McDonald's, Coca-Cola[22] and American Airlines were among 70 or so companies that endorsed the CERES Principles and pledged to go beyond the requirements of the law. By adopting these principles, they publicly affirm responsibility toward the environment, and commit themselves to release a standardized corporate environmental report every year.

The CERES Principles can have a practical application, since they establish an environmental ethic with criteria by which investors and others can assess environmental performance. These Principles, however, do not create legal liabilities, expand existing rights or obligations, waive legal defenses, or otherwise affect the legal position of any endorsing company, and are not intended to be used against an endorser in any legal proceeding for any purpose. Nonetheless, by enlisting themselves as members of CERES, companies demonstrate a beyond compliance leadership toward sustainability. By engaging in a Green Club with tough principles, companies work toward the legitimation of their efforts for more sustainable industries, and are able to point out to critics that they are doing as much as possible. As it will be discussed later in this chapter, at times they are even able to push for stricter regulations, or promote practical measurers that can enhance the eco-accountability of firms.

An example of practical measure that itself resulted in another Green Club is the Global Reporting Initiative (GRI),[23] which was born as a partnership between CERES and the United Nations Environment Program (UNEP) (Item 6 in Figure 4.2). The GRI was established in 1997 with the mission of developing globally applicable guidelines for reporting on the economic, environmental and social performance of corporations, governments and Non-governmental Organizations (NGOs). In 1999 the GRI released the first draft of the Sustainability Reporting Guidelines, which was the first global framework for comprehensive sustainability reporting. The guidelines were built with the participation of corporations, NGOs, accountancy organizations, business associations and other stakeholders from around the world. In 2002, the GRI was established as a permanent, independent, international body with a multi-stakeholder governance structure. By 2007, there were 665 organizations reporting according the GRI. Today the organization is based in Amsterdam and its core mission is maintenance, enhancement and dissemination of the guidelines through a process of ongoing consultation and stakeholder engagement.

A different kind of VEI that received a much larger business acceptance is a coalition of international private companies that advocates economic growth and sustainable development as compatible achievements. The World Business Council for Sustainable Development (WBCSD)[24] was formed in January 1995 and by 2008 its two hundred or so member companies represented 35 countries and 20 industrial sectors. The WBCSD promotes the development of closer cooperation between business, government and organizations concerned with the environment and sustainable development. The group encourages high standards through the promotion of business leadership in environmental management, the promotion of best practice demonstration programs, participation in policy development and global outreach through global networking.

The formation of the WBCSD can be seen as an industry coalition for the 1992 United Nations Conference on Environment and Development (UNCED) held in Rio de Janeiro, Brazil (also known as Earth Summit 92). At that time, industry leaders were concerned with the possibility of the UN meeting to trigger more stringent governmental regulations. Although the conference itself did not result in any major demand for better business environmental practices, most of today's initiatives were consolidated after the UN meeting, and the lasting years of the 1990s were profuse in terms of business responses to the environmental challenge.

The Global Compact is exemplary in this respect.[25] Founded in 1999 by UN agencies under the leadership of Kofi Annan, the then UN Secretary-General, civil society stakeholders and business leaders, the Global Compact is the world's largest voluntary corporate citizenship initiative. It builds on ten principles in human rights, labor law, environmental protection and anti-corruption. To become a member, companies need to make the Global Compact principles part of business strategy and operations, and facilitate cooperation among key stakeholders by promoting partnerships in support of UN goals. Although the principles are not enforced by the UN, some are enforced by national and international human rights and labour laws. Companies are required to include the principles into business strategy and report every year on the progress made.

By 2008, the UN Global Compact included around 3000 corporate participants from 116 countries. Among these companies, one-fifth was from the Financial Times Global 500 list; around half of them were small and medium sized, the other half had more than 250 employees. According to the 2007 Annual Progress survey,[26] the three major reasons for participating in Global Compact were: increased trust in company, networking opportunities and address humanitarian concerns. Overall, there was a important change of rationale for companies to participate in such VEI: "The business case for principles-based change is no longer just about avoiding costs for getting it wrong; it is increasingly about the benefits for getting it right. There is growing recognition that voluntary and regulatory efforts are complementary".[27]

Indeed, Responsible Care, a sector-specific *self regulation*, CERES, GRI and Global Compact, multi-party initiatives, and the WBCSD, an initiative of business leaders, can serve as representatives of a myriad of Green Clubs that have been used by corporations to communicate their leadership toward environmental protection. They are all voluntary initiatives that have the intention of acting as a soft regulatory measure, with the only difference that non-compliance does not result in penalization. Such feature generated mixed outcomes – from both perspectives of environmental protection and advantage for business. Before we discuss such outcomes, however, the next session presents another type of Club that emerged in the mid-1990s and became extremely relevant for business to demonstrate the environmental quality of their organizational processes.

### Process certification clubs: assuring environmental quality

ISO 14001 does more, in environmental terms, for dirtier firms than for cleaner ones[28]

While there is still limited evidence that the endorsement of generic Green Clubs, such as CERES, GRI or Global Compact bring clear reputational advantages for firms, the emergence of standards for Environmental Management Systems (EMS) caused a furor in the 1990s for the hope that they would result in competitive advantages. The two main initiatives that have been in the centre of the debate on corporate sustainability are the Eco-Management and Audit Scheme (EMAS) and the ISO 14000 series. Both initiatives are voluntary and their character of environmental auditing is intended as an internal management tool to monitor performance.

Although there are similarities, they are different in terms of legal proceedings, assurance of regulatory compliance and performance and the type of governance. The main difference, however, relates to the public participation during the process of designing EMAS and ISO standards for EMS and subsequent scrutiny of their application. (The reason for EMAS to be located under Item 7 in Figure 4.2.) EMAS resulted from a negotiation process between the European Commission (EC) with industry representatives, environmental groups and a wide array of stakeholders before it was released in June of 1993.[29] The standard is an inter-governmental initiative that has been proposed for the European Union (EU). Once the standard has been adopted by the organization, it requires compliance to the environmental laws of the country where the company operates, and third-party verification by an accredited EMAS auditor. Overall, the EMAS is mostly confined to the EU.

The ISO 14000 series follow a different trajectory. The global profile of the previous success of the ISO 9000 series of Total Quality Management (TQM) was the main cause for the expectation that environmental certification would eventually result in competitive advantage for firms. Having emerged in the 1980s, the TQM move-ment has highlighted the influence organizational processes exert on the overall competitiveness of the firm.[30] During the 1980s and 1990s, quality-oriented management has enhanced the competitiveness of an impressive number of corporations worldwide. By identifying the ultimate sources of quality problems, firms reduced or eliminated trade-offs between costs and quality. Companies that pursued zero defects and continuous improvements of organizational processes have improved the quality of products and services at the same time that costs were reduced. Since products and services of enhanced quality (and reduced costs) have a better chance to succeed in the marketplace, such practices obviously influenced their competitiveness.

Many hoped that the environment-related ISO 14000 series would generate similar results for companies of the 9000 series on quality. However, as it was outlined in Chapter 1, there is a fundamental differ-ence between quality and environment that can limit the possibility of some gains to be valued by clients. Quality improvements can be transferred from organizational processes directly to the products and services bought by consumers. This embededdness allows quality to become a private profit. Environmental protection, on the other hand, is a public good and therefore cannot be (directly) transferred to prod-ucts and services. From the point of view of consumer, environmental protection is about harming (or being harmed) less, rather than receiv-ing more for a purchase. Those who expected that the ISO 14000 series would have the same effect the ISO 9000 series had on firms' competitiveness fail to recognize this fundamental difference.

Does it mean that the certification of EMS cannot be used to dif-ferentiate firms or become a source of competitive advantage? Not at all, but because the levels of economic benefits depend on a wide array of variables – ranging from internal capabilities to the structure of the industry – only particular conditions favor firms to gain such advan-tages. Once again, the question is not whether it can work but when. While the value of quality is more immediately measured through the intrinsic characteristics or performance of products and services, the value of environmental and social responsibility is less objective and depends on a broader set of consumer perceptions. What eventually counterbalances this argument is the fact that some environmen-tal aspects have gradually assumed a quality status, increasing the importance intangibles have in the success of businesses.[31] Under par-ticular circumstances, consumers and client organizations increasingly attribute some value to the way organizations manage their production processes and supporting activities.[32] For instance, when automakers Ford, General Motors and Toyota announced in 1999 that they would require their suppliers to certify their EMS according to ISO 14001, the first ones to obtain certification certainly had an advantage.[33] For the automakers, the EMS certification could indicate good housekeeping and high quality of the operations. Hence, environmental prerogatives in this case have a close association with the overall quality of organizational processes. For companies supplying products or services to other corporations (also known as industrial markets or business-to-business – B2B), this has practical value.

In the case of consumer markets, things are a bit different. Observe the case of Pallister, a winemaker from New Zealand (NZ) who was the first winery worldwide to obtain an ISO 14001 certification in 1998. Since its foundation in 1982, Pallister Estate Wines has invested heavily in product quality and aims at becoming recognized as a leader in the industry so that it can attract high-end customers. Pallister wines appeal to about 5 percent of the NZ wine drinkers and the company exports 60–80 percent of their production. The winery saw its ISO 14001 certification as part of their commitment to environmental responsibility and high-quality wines, which is appreciated by customers. But the management doubts that customers are willing to pay for environmentally responsible wine in the same way they are willing to pay for other traditional attributes, such as vintage and particular mixes of grapes.[34]

On the other side of the globe, in Brazil, a local company operating in the cosmetics industry has been investing in corporate social and environmental responsibility for two decades.[35] The Boticário has long been committed to cleaner technologies, the development of recycling programs and the reduction of waste generated by its packaging. The company has also been sponsoring social projects in local communities as well as conservation projects through its Foundation.[36] Similar to the NZ wine company, though, consumers of Boticário established little connection between what the company does and what it sells.[37] Although they are sympathetic to the company's green values, which are reflected in the fabrication processes, product convenience and fashion trends are the main determinants of purchasing. Similar to wine consumption, for most consumer goods, eco-management excellence of organizational processes is a plus; it is rarely the main factor determining the purchase.

A final and more personal way of thinking about the importance of ISO-certified EMS for final users is to put your consumer hat on: For instance, how often do you choose a hotel because of its environmental credentials? When choosing a hotel for holidays, do you evaluate whether the hotel is certified according to ISO 14001? More importantly, would you decline checking in a hotel just because it does not have a certified EMS? Although over time more people may ask for such things, today this number is certainly very low. Large corporate buyers, on the other hand, are increasingly demanding environmental credentials from their suppliers – including hotels – but, as the next chapter explores in more detail, individual consumers have limited information or too many countervailing variables to consider when shopping. As a result, the eco-performance of organizational processes is hardly a consideration in their purchase.

What about the new ISO 26000 standards on social responsibility (launched in 2009)? Can we expect it will be substantially different from the ISO 14000 series? The new standard provides guidance for Corporate Social Responsibility (CSR), facilitates interaction between business and society and provides guidance on how organizations can operate in a socially responsible way. ISO has established Memorandums of Understanding with the International Labor Organization and UN Global Compact in order to cover human rights, labor, environment and anti-corruption issues. Compared with the development of the 14000 series, there was a much wider stakeholder consultation in the 26000 series: besides industry and government, there were also representatives of NGOs, consumers, unions, academia, including geographical and gender-based balance of participants (Item 7 in Figure 4.2). Although it is too early to evaluate the benefits firms may gain from certifying CSR activities via ISO 26000, we cannot automatically assume that it will generate advantages for the early adopters. The standards have been efficient in shielding companies from bad reputations, but the possibility of generating advantages will depend on the context in which they are deployed. The next session presents another special type of Green Club that emerged in the 2000s and became extremely relevant for business to demonstrate beyond compliance leadership.

### *Climate clubs*: building good reputation?

While the 1990s were marked by the emergence of generic and process certification (Green) Clubs, such as the ones described in the previous section, several Climate (change) Clubs were founded in the early years of the new century. In other words, the broad scope of sustainability issues eventually became more specialized or, in the eyes of some environmentalists, reduced to the issue of greenhouse gas emissions.

The Business Environmental Leadership Council (BELC) was one of the first initiatives, created in 1998 as a specialized program of the Pew Center for Climate Change. The Pew Center carries out research on climate change policy but membership is free of charge. The center helps companies to implement climate change strategies and to seek to influence policy and regulation in this area. BELC is the largest US-based climate change association including 42 members, which are mostly from the Fortune 500 and come from high-tech industries, manufacturing, chemicals and utilities. They work with the center to develop practical solutions and formulating policy members are expected to implement proactive and innovative measures including: setting targets for Greenhouse Gasses (GHG) emissions reductions; implementing innovative energy supply and demand solutions; improving waste management practices; participating in emissions trading; and investing in carbon sequestration opportunities and research.[38]

Climate Savers is another initiative created in 2000 by the World Wildlife Fund (WWF) in cooperation with a handful of major firms, such as Johnson & Johnson, Lafarge, Nike, Novo Nordisk, Polaroid, HP, IBM, Nokia and Tetra Pak. The goal is to establish ambitious targets to reduce greenhouse gas emissions on a voluntary basis. Partnering companies have each developed a business plan in cooperation with the WWF on how to cut emissions and work with climate change issues in their businesses with regard to their main business activity. One major condition to join the Climate Savers partnership is that the target cuts agreed with WWF must be more ambitious than previously planned by the company. Activities that could be included in the plan are: transport efficiency, energy-saving products, energy efficiency and fuel switching.[39] Collectively companies have agreed to reduce carbon emissions by 13 million tons annually by 2010. Partnering companies have reported that they have increased efficiency and saved hundreds of millions of dollars.[40] Catalyst, a US-based Information Technology (IT) company, for instance, has promised to reduce its GHG emissions by 70 percent by 2010, over 1990 level emissions. Outside experts monitor and verify the company's compliance with the agreement. The more ambitious participants also try to influence the carbon emissions of their suppliers. WWF offers companies networking opportunities and guidance/best practices in developing accounting and mitigation plans. Nike, a Climate Saver partner, finds that the goals of the group help them move ahead of competition and innovate.

According to Sarah Severn, Nike's Director of Corporate Responsibility Horizons:

> Participation in Climate Savers enabled us to get an early start on an issue that has major consequences for business and society. We have found that constraints can lead to tremendous innovation and despite growth in our owned and managed operations we have become more efficient with our energy use. Our next steps will be partnering with suppliers to further reduce our manufacturing and logistics climate footprint.[41]

The American NGO Environmental Defense launched another Climate Club in October 2000: the Partnership for Climate Action (PCA). The PCA is a small group formed by a few companies operating in the energy, oil and gas businesses: Alcan, British Petroleum (BP), DuPont, Entergy, Ontario Power Generation, Pechiney, Shell International and Suncor (observe that Suncor was the main partner of the Stuart Oil Shale project in Australia, mentioned at the beginning of this chapter). PCA is dedicated to climate protection, and its members are committed to sharing their knowledge, tools and experience with other companies seeking the same goals. The PCA member companies commit to declare GHG emissions limits and put management and policies in place to implement and measure the results. As an example of commitment, BP agreed to cut emissions from its own operations by 10 percent from 1990 levels by 2010, a goal that was already achieved in 2002. Innovation in this case helped BP to achieve their goal eight years early and at no net cost to the company.[42]

The US EPA founded Climate Leaders in 2002 in partnership with industry. Companies that join the program have to commit to reducing GHG emissions. The Climate Leaders helps member companies to develop these plans by completing an inventory of GHG emissions, setting aggressive reduction goals, and annually reporting their progress to EPA. In return, the companies that join Climate Leaders get to work with the group to develop comprehensive climate change strategies. The Climate Leaders have 153 partner companies; 50 percent of them are listed on the Fortune 500 list. However, by 2008 only 80 companies had announced their GHG reduction goals and only 11 had actually achieved the goals. EPA publishes a list of the companies who have achieved their goals, the ones who have set targets and the ones who have goals under development but that are not yet met.[43] Through program participation, companies create a credible record of their accomplishments and receive EPA recognition as corporate environmental leaders.

Finally, the Climate Group was founded in 2004 as an "independent, non-profit organization dedicated to advancing business and government leadership on climate change".[44] Members include some major corporations such as BP, HSBC, Virgin, and Bloomberg, various national and regional governments, such as the cities of New York and London, the state of California, British Colombia, Ontario, Quebec and the Australian states of South Australia and Victoria. Members are requested to pay a fee and facilitate matched funding from charitable trusts. Apart from committing to a GHG reduction plan and reporting it, which have to be endorsed by top management, member companies also commit to sharing knowledge.

The Climate Clubs mentioned above are representative of the main initiatives, but the list is certainly not exhaustive. Overall, almost all initiatives offer expert advice, research and consultancy to help their member companies set up climate change plans and do their GHG emission accounting. Some Climate Clubs also offer publicity for their members. With the exception of Climate Group and WWF Climate Savers, the others do not accept money or funding. A typical mission statement is: "We promote the development and sharing of expertise on how business and government can lead the way towards a low carbon economy whilst boosting profitability and competitiveness" (Climate Group). In return, they ask club members to develop a climate change strategy, reduce emissions aggressively, share knowledge and best practices and apply innovative new business models to climate change issues.

Most Climate Clubs are based in the US, which might reflect the greater need for voluntary initiatives in the absence of government regulation, and the fact that the Bush government did not ratified the Kyoto protocol. Many companies such as Shell, DuPont, BP, and Johnson & Johnson are partners of several climate change initiatives. These are mostly companies that have high levels of GHG emissions and therefore are targeted by eco-activists. In fact, Shell and DuPont went further and joined another club with a different mandate: lobbying toward stricter $CO_2$ emission regulations.

The politically oriented Climate Club was formed in 2007 by businesses and leading environmental organizations[45] to call on the American federal government to enact strong national legislation for the reduction of GHG. The United States Climate Action Partnership (USCAP) has issued a set of principles and recommendations to underscore the urgent need for a policy framework on climate change. The lobbying group offered a set of recommendations for the general structure and key elements of climate protection legislation. The Congress should specify, for instance, an emission-target zone aimed at reducing emissions by 60–80 percent from current levels by 2050. The USCAP also asked for legislation that requires actions to be implemented on a fast track while a *cap and trade* program is put in place. This includes the establishment of a GHG inventory and registry, credit for early action, aggressive technology Research and Development (R&D), and policies to discourage new investments in high-emitting facilities and accelerate deployment of zero and low-emitting technologies and energy efficiency. Finally, the group also lobbied to reward those firms that have acted to reduce GHG emissions and encourage others to do so while the program is being established. In the opinion of Andrew Hoffman,[46] this should not surprise anyone:

> When the companies of United States Climate Action Partnership (USCAP) – businesses including GE, Alcoa, DuPont, and PG&E – announced their call for federal standards on greenhouse gas emissions in January 2007, the wall street journal castigated these "jolly green giants" for acting in their own self-interest in promoting a regulatory program "designed to financially reward companies that reduce $CO_2$ emissions, and punish those that don't". But seeking advantage is what companies do. Any company that can foresee business opportunities in influencing carbon-emissions regulation is practicing what is expected of business managers – capitalism.

Hoffman brings back the recommendation made by Michael Porter for beyond compliance leaders to lobby for stricter environmental regulations.[47] The rationale is simple. Since the return on eco-investments depends on the "rules of the game", for large emitters such as Alcoa, Shell and DuPont, uncertainty about GHG regulations is far worse than a reasonable (negotiated) regulatory regime. Investments in carbon sequestration, for instance, only make business sense in a regime in which there is a *cap* for emissions – as it was discussed in Chapter 3. In the absence of a national regulation, as it has been so far in most places outside the EU, such *cap* can emerge as a self-imposed regulation via Climate Clubs. However, putting into place a credible self-regulation scheme can be expensive and may take time because gaining institutional legitimacy for the Club requires acceptance of the standards by relevant stakeholders. In order to reduce the risk of rejection, the formation of some Climate Clubs involve a wide range of stakeholders (see Item 7 in Figure 4.2), making the formative stages lengthy due to demanding negotiations. It is only natural, then, that club founders will lobby governments to transform self-regulation into law. By pushing the frontiers of national and international regulatory regimes, club founders will try to extract rents from their investments, since membership and compliance will be more costly for late entrants.

Club setters will also try to obtain reputational benefits for being identified as leaders of self-restraining initiatives to reduce emissions beyond legal obligation. Nonetheless, due to the public uncertainty about the issue, gaining reputational advantage from climate change may prove much harder than other environmental issues.[48] Independently of the scientific evidence that the planet is warming,[49] as well as the plethora of negative climactic consequences, there is still a large portion of academia, business and general public that simply does not accept it as an issue to be urgently addressed.[50] In this respect, Climate Clubs do imply a proactive role of large multinationals. Considering most Clubs have been formed in times of high uncertainty, they suggest many large corporations learnt from past mistakes.

Different from the early days of Green Clubs, which emerged out of crisis such as the Bhopal and Exxon Valdez accidents, businesses have tried to anticipate stakeholder pressure in a better fashion they did previously by proposing self-regulations for GHG ahead of governments. Even though uncertainties about climate change may remain, by being more open about the problems they may eventually have to curb, many MNCs have been trying to walk the talk. Overall, rather than just avoiding bad reputation, by proposing beyond compliance standards via Climate Clubs, they are working toward building positive reputation. Whether Climate Change clubs will succeed remains an open question. Nonetheless, the recent experience of other Green Clubs, such as Responsible Care, GRI and ISO 14001 standards suggest that, in the long term, they can be instrumental in helping companies to improve their reputation. Therefore, it is opportune to inquiry into the reputation of Green Clubs themselves: What factors influence their reputational value?

### The reputational value of Green Clubs

By becoming members of Green Clubs corporations presumably will improve their environmental performance. The standards of Responsible Care, for instance, must be translated into a number of operational areas, with measurable targets set for performance improvement. Similarly, the commitment to CERES Principles can trigger organizational innovation and generate better corporate reputation. The improvements, however, depend on the constituency of Green Clubs:[51] often, the more demanding clubs were founded by a civil society organization, which aimed at gathering a wide range of stakeholders. Sector-specific Green Clubs, on the other hand, tend to be initiated by industrial associations with the original aim of protecting members against bad reputation. At least in the early stages, these clubs are inclined to concentrate on legal compliance and cost-cutting exercises, as the Responsible Care did for the chemical industry. CERES, on the other hand, was founded by a civil society organization and, as a result, its requirements are more demanding. This explains the lower number of companies that endorse its principles, when compared to the members of the World Business Council for Sustainable Development (WBCSD), which is a business initiative, or to the Global Compact, which is quite loose in terms of the implementation of its principles.

Overall, from the environmentalist's lenses, the proliferation of Green Clubs with their general principles, codes of conduct and behavior guidelines is good news; they influence businesses to lower their impacts on nature. However, business people may wonder: by joining such clubs, do companies gain any advantages? Do they improve the reputation and generate eco-differentiation for the members? Although it is not yet possible to have definitive answers, two aspects have already emerged. First, these initiatives have served defensive goals very well. In the majority of cases, by endorsing VEI companies have avoided the disadvantages of being exposed to bad reputational campaigns. For instance, after the problems Shell faced in Nigeria and with the Brent Spar platform in the North Sea, the company made efforts to legitimize its efforts by joining several Green Clubs, such as Responsible Care, Global Compact, and Corporate Leaders Group on Climate Change. Shell also started using sustainability reporting as a key communication tool. Although the controversies surrounding its operations never vanished, by pursuing a consistent strategy based on beyond compliance leadership, the company managed to protect its reputation from further harm.

Second, it is important to reemphasize that the reputation of the Club influences the possibility of a company to increase its own reputation. Clubs formed by a wide range of stakeholders (Item 7 in Figure 4.2), which require demanding standards from members have higher reputational value than the ones formed by business alone, or the ones with relaxed requirements. Besides high improvement requirements for members, the reputational value emerges from the institutional legitimacy obtained via stakeholder dialogue during the formation of the club. Since Green Clubs intend to form a credible self-regulation scheme that is accepted by relevant stakeholders, in order to manage the non-acceptance risk the club's initiators often involve as many stakeholders as possible. As a result, during formative stages negotiations of multi-stakeholder clubs are lengthy and eventually costly. Once the Club is formed, however, stakeholder involvement will confer decisive legitimacy for positive reputation.

Besides influencing the reputational value of the club, being a founder can bring other benefits. If the club's standards are accepted by stakeholders, by participating in the early stages, the company will be able to have lower costs than late entrants, since laggards will need to comply with the demanding standards set by the leaders. This has, if fact, been the prescription of Forest Reinhardt[52] to manage competitors. By raising the bar of environmental compliance via private regulation (set by Green Clubs), leaders can gain competitive advantages. Besides lower compliance costs, the early mover status will also provide the company with the reputational benefits of an environmental leader. There is a drawback, however. Leadership will increase the public exposure quite substantially, requiring the company prepare itself for wider public scrutiny.[53] Hence, in the medium to long terms, leadership may pay off in economic and reputational terms, but companies will have to be able build competencies in stakeholder dialogue and learn effective ways of communicating with the wider public.

## WHEN *BEYOND COMPLIANCE LEADERSHIP* PAYS

The adoption of Green Club's doctrines and guidelines, such as CERES Principles, Global Compact or the GRI may help corporations influence a positive public opinion about organizational practices, and eventually enhance their reputation. Although relevant, such knowledge is simply too broad. Managers need more precision about the real outcomes of using Green Clubs as vehicles to construct positive reputation. Overall, they need to know when these types of investments pay off or lead to competitive advantage. Unfortunately, there is no silver bullet for such question. As it was emphasized earlier, the return on eco-investments depends on type, timing, context, as well as the evaluation method of the investments, which can be based on tangible and intangible assets. Broadly, these conditions relate to the overall context in which the company operates, the constituency of Green Clubs, as well as the type of consumers served by the company. The circumstances found in consumer markets (B2C), for instance, normally differ from those of industrial markets (B2B), and differentiation strategies developed for one context cannot be transplanted to the other.

## Context

As the cases of Suncor and Shell suggest, beyond compliance of organizational processes will benefit MNC that are susceptible to the pressure of local stakeholders, which can mobilize public opinion in their home base. Stakeholder response to organizational practices is prone to happen when an issue is created around a specific concern, so the public is sensitized. As it happened in the case of Shell, such mobilization can occur thousands of miles away from the industrial site. Such conclusions are in line with the findings of a previous study,[54] which identified a significant positive relationship between the market value of MNC and the level of environmental standards used in their factories worldwide. Having homogeneous beyond compliance standards across plants, besides being cost effective, in this case is also valued by the market. Although there are many speculative reasons behind these findings, lower reputational risk is possibly among them.

Some firms with highly visible brands operating in consumer markets (B2C), which depend deeply on natural resources, can also benefit from Strategy 2. Some of them have also been increasingly targeted by eco-activists and, in order to avoid reputational risks, have good reasons to develop excellent stakeholder dialogue. Coca-Cola is an informative example. Although the company has been a leader in corporate environmentalism, such as water recycling in its plants and post-consumption recycling schemes, it has not branded itself on the basis of environmental prerogatives.[55] There are good reasons for this. For a company that reaches one billion people a day with its various brands of soft drinks, bottled water and juices, it would certainly be too risky to explore eco-branding strategies (analyzed in detail in Chapter 5). As the number one brand in the world,[56] Coca-Cola has correctly avoided linking the brands of its products to environmentalism, and rather kept a relative low profile focusing on Eco-efficiency (Strategy 1).

More recently, however, Coke's reputation has been increasingly subjected to anti-globalization critics, forcing the company to refocus its sustainability strategy. As *Fortune* magazine[57] put it: "There are solid business reasons Coke wants to become more sustainable. The company came under attack when wells run dry near one of its bottling plants in India. And bottled water like Coke's Dasani has become a hot-button issue for activists". Therefore, stakeholder dialogue for the (negotiated) sustainable management of local aquifers is becoming increasingly important for the company to guarantee its license to operate. Coca-Cola certainly has enough money to pay for the water it needs, but in many parts of the world, obtaining water will increasingly depend on stakeholder negotiation, rather on capital alone. Eco-activists will be ever demanding on Coke to demonstrate leadership in the way the aquifers are managed. Coca-Cola has good reasons to tell stakeholders that it follows the principles of CERES and Global Compact of social and environmental responsibility, and has most of its factories certified according to the ISO 14001. They are key instruments for Coke to avoid bad reputation and facilitate communication with stakeholders.

Producers of commodities such as minerals, oil and gas are in a different league. They have very little scope to differentiate their non-renewable hydrocarbons; much less to brand products on the basis of ecological prerogatives. However, they share one central aspect with Coca-Cola. These firms have been under increasing pressure from key stakeholders, and have sound rationales to focus on the excellence of their organizational processes. Besides being commodities (i.e., common to any producer in the industry, regardless its effort to be unique), hydrocarbons are non-renewable and tend to be perceived as eco-unfriendly.[58] Hence, for these companies, focusing on process-oriented beyond compliance is the most logic and reasonable way of demonstrating environmental leadership.

Firms supplying other businesses (B2B), which are themselves under pressure to improve their environmental performance, also benefit from Beyond Compliance Leadership strategies. ISO 14001 certification, for instance, have a clear value for the client organization because it makes easier for suppliers to communicate to potential buyers that they manage operations according to standards of best practice.[59] Using an academic terminology, ISO certification reduces information asymmetries throughout the supply chain.[60] When automakers Ford,

General Motors (GM) and Toyota announced in 1999 that they would require their suppliers to certify their EMS according to ISO 14001, the first ones to obtain certification gained an obvious advantage.[61] It is also true that the EMS certification represented a first-mover advantage for a short period; by 2002 it became a mere license to operate in the industry. This is not, however, unique to strategic environmentalism. As in almost every sphere of management, competitive advantage is obtained in a relatively short window of opportunity, which is open only for the ones who are ready at the right time; the ones who are in the forefront of a specific management practice. As firms within an industry adopt more ambitious proactive practices, the beyond compliance frontier moves further, and what once was a differentiator (such as a certified EMS) becomes a normal, non-competitive practice. In Scandinavia, for instance, where practically all dairy producers hold ISO 14001 certifications, beyond compliance is moving toward more demanding issues, such as voluntary standards for animal rights. There, beyond compliance leadership already requires firms to develop competences in animal welfare and ethics.

The globalization of supply chains is a trend that partly explains the dissemination of ISO standards.[62] When buyers and suppliers are distant from each other, the value of certification is instrumental; more so for companies located in emerging economies. Today the standards of quality in some sectors are practically independent of location, and suppliers presenting the highest degree of environmental performance in their operations make a good point to get contracts. In sectors such as garment and footwear, stakeholder pressure extended the environmental responsibility to the global supply chain. The reputation of the buyer organization, which is normally located in a rich country, can be affected by the upstream practices throughout the value chain in emerging economies, such as Vietnam or Mexico. Responsible operations throughout the value chain become crucial. For instance, after being accused of having sweatshops throughout their supply chain, Nike and Levi Straus imposed strict social and environmental standards to their suppliers worldwide. Such standards create a non-tariff barrier to the suppliers, which have to make their best to prove their beyond compliance behavior. Leadership, in this case, certainly pays off for the suppliers. In the optic of export-oriented companies located in countries such as China, India and Mexico, using certified standards as backbones of their beyond compliance strategies can certainly generate advantages.[63] Indeed, the number of certification in developing countries indicates this value.[64]

## Competences

In order to choose the adequate sustainability strategy, managers should ask themselves whether the business would suffer a blow if stakeholders judged the company to be promoting inadequate green image. They should also have a clear understanding of what it means to be green in the industry.[65] In order to deploy strategies based on beyond compliance leadership, they should have competences that match their ambitions. The case of BP is exemplary in showing how a not carefully chosen sustainability strategy may damage corporate reputation. By spending US$200 million in a public relations campaign to change from British Petroleum to *Beyond Petroleum*, BP indicated its intentions to become a beyond compliance leader, provoking high expectations among eco-activists, consumers and the general public. However, in the years following the campaign, the investments to go beyond petroleum were minimal when compared with the ones to remain in the petroleum industry. As stakeholders realized that the company would not be able to move away from its core product (oil) in any time soon, frustration built up. Most possibly, BP executives knew that moving beyond petroleum would be impossible in at least one generation (25 years), and so become the beyond compliance leader in the industry (Strategy 2), would require much more effort and resources. By suggesting that such move would happen sooner, BP shot itself on the foot. On the other hand, BP has been extremely successful with its internal emissions trading schemes, which resulted in more than US$650 million in savings. With such type of performance and the awareness that oil is going to be the mainstream activity of the company for the time being, BP should refrain from claiming beyond compliance leadership and rather focus on Eco-efficiency (Strategy 1), which could render safer outcomes.[66]

As the example suggests, environmental reputation requires efforts to communicate their beyond compliance leadership to the general public, but it is fundamental that companies do their homework first. In order to deploy beyond compliance leadership successfully, managers need to address the following questions:

- Do we know what is our reputation regarding environmental and social issues? How do we measure it? How reliable is our evaluation?
- What are the main conflicting stakeholder interests in our business? What are the most influential stakeholder groups demanding actions form the company? Whom should we prioritize and why?

- How prepared are we to respond to eco-activism? Are we able to communicate openly with radical eco-activist groups? Are we prepared to satisfy their demands?
- Are we communicating our sustainability efforts effectively? How is our relationship with local communities? Do they value our efforts to reduce our impacts? How do we know it?
- Are our business clients using positive or negative screening to evaluate us? Do we run the risk of loosing contracts because of the environmental performance of our processes? If so, what should we do about it?
- Should we invest on the implementation and certification of an Environmental Management System (EMS)? Do our competitors have it? Has ISO 14001 certification become a license to operate in the industry? If so, what should we do in case we decide to be the beyond compliance leader in our industry?
- Based on what should we join any specific Green Club? Would our reputation be positively affected by our membership? Why? Can a Green Club membership help us to become a beyond compliance leader?
- Should we make a clear declaration about any emission reductions targets? Do we have the internal competences to achieve them? Do we need help?
- Can we realistically be the beyond compliance leader in our industry? Can we beat competitors in any area of environmental leadership, or we run the risk of frustrating our stakeholders?

## CONCLUSION

This chapter explored the possibility and, for some companies, the need to focus on process-oriented differentiation strategies. Companies that invest on Beyond Compliance Leadership (Strategy 2) will do as much as they can to improve the efficiency of their operations, but because the main issue they face is reputational, they need to make beyond compliance the focus of a differentiation strategy. They need to protect their reputation against eco-activism by investing in stakeholder dialogue and engagement. Besides doing their homework and reducing their environmental impact, many have engaged in Voluntary Environmental Initiatives (VEI) or Green Clubs, as they are also called, as a means by which they can legitimize and promote these efforts to key stakeholders and the wider public. The expansion of such initiatives, which fundamentally are non-state regulations, so-called *civil*, *private* or *soft* regulation, suggests the failure of existing institutions to regulate firms at global level. This poses a basic question for business: how effective is this new mechanism in terms of business profits, when compared with public benefits? When is civil regulation a differentiator?

The chapter provides insights into these questions by critically reviewing the most prominent Green Clubs, and verifying whether they helped firms to protect or enhance their reputations. Overall, we can safely say that they protected many companies against bad reputation. Responsible Care and Coalition for Environmental Responsible Economics (CERES) are good examples. More difficult, however, is to identify generic differential advantages provided by these clubs. Although some companies gained reputational advantages, club membership by itself does not bring that. Considering the differentiation depends on reputational asymmetries, club membership *per se* is not a guarantee of value creation. ISO 14001 certification, for instance, does not automatically enhance firm's reputation. Once again, the context in which the company operates is vital for the success of a process-oriented differentiation strategy. The level of diffusion of the standard within a specific sector influences its chance to bring competitive advantages.

Overall, in order to obtain differential advantages from process-oriented strategies, managers have to identify who is valuing the environmental investments of the firm. They need to ask: which (group of) stakeholders praises the company's efforts to, for instance, comply with the Global Compact principles, certify its Environmental Management Systems (EMS) or produce sustainability reports according to the Global Reporting Initiative (GRI) guidelines? If so, what are the criteria? Some investors, for instance, have been using the Dow Jones Sustainability Index (DJSI) for such purposes.[67] Hence, if investors are the key stakeholders to be pleased, then their evaluation through the DJSI is good enough. But more demanding stakeholders do not see such indexes as reliable indicators of sustainability, and even though companies would have good reasons to be among the top DJSI performers, they should know that the Index will not protect them from bad reputation. They should be prepared, for instance, to be criticized for supplying the data themselves and not be audited.[68] If the key stakeholders are eco-activists, then close stakeholder dialogue and engagement is more effective than just generic rankings, which often are used as additional risk management criteria for investors.

# 5

# ECO-BRANDING

The prospect of differentiating products and services has always appealed to business. After all, who would not want to have loyal customers who pay premium prices so the company can avoid razor-thin margins of price-sensitive markets? Higher margins allow the company to better absorb market fluctuations, invest in R&D, among many other good things brought by wealthy customers. In order to differentiate from competitors, firms try to provide something unique that buyers value beyond the price.[1] In one way or another, all companies try to do this – even when competing on the bases of low price. They try to be unique so that their clients will prefer them. However, success depends less on the efforts to be different than on how consumers value these efforts and, more importantly, whether they are willing to pay for it.

What about differentiation based on ecological prerogatives? Is it any different from traditional differentiation strategies? If so, what is different? As this chapter delves into, there are both subtle similarities and differences between traditional and ecology-oriented marketing strategies for products and services. On the resemblance side, historical data suggests that ecological differentiation tends to be restricted to (relatively) small market niches. This should not be a surprise. Only a minority of consumers is wealthy enough and willing to pay price premium for both high-end and eco-oriented brands. Indeed, today eco-oriented products represent a defined market niche explored by a large number of firms worldwide. In this respect, marketing differentiation based on environmental attributes can be treated similarly to traditional niche market strategies.

There are, nonetheless, crucial differences. Eco-differentiation imposes additional requirements to products: They need to present lower environmental impacts than similar products while satisfying all other requirements, such as quality, convenience and aesthetics. In other words, it is not only about private benefits for the consumers (and private profits for the company), but also about the public benefits embedded in the product. The treatment of such public benefits is what distinguishes eco-differentiation from traditional-differentiation strategies. First, there is the issue of credible information. For instance, the activity system involved in producing a dining table needs to be clearly displayed to consumers. They have to know about public benefits, such as biodiversity preservation during tree harvesting, water recycling during production process and the minimization of carbon emissions during transportation. They also need to trust the methods used to assess data, as well as the source of information.

One way of presenting such information is via eco-labels. Companies use eco-labels as a means of simplifying the information embedded in their products, legitimize their efforts and, more importantly, use them as central components of eco-branding strategies – discussed in detail in the last part of the chapter. Is opportune to ask then: when are eco-labels effective differentiators? When certified eco-labels pay off? What kind of products and/or contexts eco-labeling can generate competitive advantage? When are consumers willing to pay for eco-labeled products? How can the success of an eco-label scheme be measured? This chapter addresses such questions in a period in which companies such as Wall Mart and Tesco are proposing climate (or carbon) labels for the products they retail. Hence, the chapter discusses the potential of carbon labels to deliver what was expected from other eco-labels, such as the German Blue Angel and the European Union (EU) Flower. Is the success of eco-labeling schemes compatible with eco-branding of individual companies? The case of Forest Stewardship Council (FSC) is exemplary to open the discussion.

## PRODUCT CERTIFICATION CLUBS: ECO-LABELS

The modern sustainable forestry movement emerged in the early 1980s, as concerns about the state of world's forests and how they were managed escalated.[2] There was a widespread frustration about the incapacity of corporations and eco-activists to agree on forestry standards and several initiatives did not succeed in reducing the pace of destruction of the world's native forests. By the end of the decade, eco-activists' campaigns by the World Wildlife Fund (WWF) and Greenpeace, among others, resulted in several retailers boycotting tropical and old growth timber. In Europe, British, German and Dutch environmentalists led the boycott against imports of tropical timber from Malaysia and Borneo. The unsustainable forestry management practices were highlighted once again during the 1992 Earth Summit in Rio, which furthered the debate and called for international standards. Retailers in Europe got together to demand certified wood/products. In Holland, 95 percent of the retailers agreed to import timber only from sustainably managed forests. In the United States (US), on the other hand, the key player forcing changes in forestry practices was the Rainforest Action Network (RAN), an eco-activist Non-governmental Organization (NGO). Even though eco-activists campained fiercely for sustainable practices in the industry, at that time there was very little direct consumer demand for certified wood.

The foundation in 1993 of the Forest Stewardship Council (FSC) is a result of this call for an international standard. The FSC began operations in 1996 with its headquarters in Oaxaca, Mexico – a location that would suggest a symbolic meaning to the cooperation between the north and the south. Efforts were made to keep the process as transparent as possible and include all stakeholders with a say in the industry. This openness provided legitimacy to the standard, which includes both environmental aspects, such as soil and eco-system protection in plantations, as well as social aspects, such as the respect of indigenous populations and labor standards. The FSC then monitors the compliance of these standards for companies that would like to apply for the FSC label.[3] Hence, the FSC does the standardization and accreditation of forestry practices.

The forest industry's first reaction to the FSC was that no certification was needed and that it would damage business. After realizing that it was an irreversible movement, in 1995 the American Forest and Paper Association (AFPA) set up its own certification scheme – the Sustainable Forest Initiative (SFI).[4] Clearly, the SFI was founded with a defensive strategy to pre-empt the more stringent (private) regulations of the FSC and avoid bad reputation.[5] Not surprisingly, environmentalists criticized the SFI for being self-serving and not sufficiently demanding. The industry, on the other hand, argued that there was no clear demand for certification from end-customers to justify the creation of a multiparty initiative such as the FSC. Many forestry enterprises in the US also supported the SFI, instead of the FSC. They were apparently right in considering that, together, the principles of SFI and standards such as the International Organization for Standardization (ISO) 14001 would satisfy retailers' demands.[6] The private regulations of Green Clubs were sufficient to protect them against bad reputation. As a result, certification schemes flourished in the first decade of the

21st century. By 2008 there were more than 50 forest certifications schemes in the world, which directly or indirectly compete with the FSC, as the SFI does. Nonetheless, the FSC stands apart, for it is the only standard-setting organization for responsible forest management supported by both the corporate and the environmental communities. After overcoming the lack of funding and organizational problems in the late 1990s, the FSC received industry recognition in the 2000s when 80 percent of the roundwood industry recognized the need for third party certification.[7] In 2005, FSC-certified companies reported an estimated market value for the labeled wood at around US$5 billion. Based on a 2007 survey, the value of FSC-labeled sales of wood products is estimated at US$20 billion. Companies with a combined estimated turnover of US$250 billion in wood products are committed to FSC certification.[8]

By the mid-2000s, about 30 percent of the forests were certified in Europe, and FSC-certified wood and wood products have around five percent share in the market. In the US, the share was much smaller. Only seven percent of the forests were certified and FSC wood and wood products accounted for only one percent of total sales.[9] In other words, the limited diffusion of the FSC label means that the public benefits of sustainable forestry are still small. From the strict sense of product differentiation, such reality would not necessarily be a bad news for business. After all, differentiation is about uniqueness, and the limited share of FSC products in the market could eventually benefit differentiation strategies. However, the proliferation of schemes and labels has worked against product differentiation. More importantly, the final consumers are simply not willing to pay a price premium for certified wood or wood products; they are not even willing to privilege stores that sell FSC wood. If the costs of certification cannot be passed onto consumers, what is the real value of the FSC label? Can FSC labeling eventually help companies to build eco-brands?

## Eco-differentiation standards and strategies

What is the fundamental rationale of the case? The FSC establishes a direct relationship between the upstream activities with the final products sold by corporations. The FSC label facilitates client organizations and final consumers to identify products of sustainable forestry practices. It is a result of multi-stakeholder dialogue involving civil society organizations such as WWF, and forestry companies, retailers and general public, which confer high levels of legitimacy and

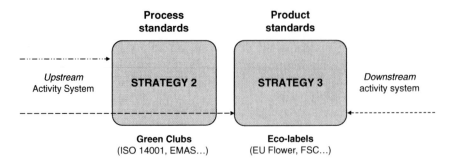

FIGURE 5.1 **Eco-differentiation Standards and Strategies**

acceptance of the standard/label by all parties. In essence, FSC is an additional example of Green Club, presented in the previous chapter. Green Clubs are instruments available to firms to protect their reputation against eco-activism, and to legitimize their efforts toward environmental protection. ISO 14001, Responsible Care or Carbon Savers, for instance, are examples of process-oriented standards. As Figure 5.1 depicts, these standards mainly address upstream activities and industrial processes, and *stop* at the organization – the reason they cannot be associated with the products sold by the certified company. On the other hand, eco-labels such as FSC are product-oriented standards, designed to be directly associated with products (and sometimes services) via direct product labeling. Although eco-labels relate mostly with upstream activities, often they also involve downstream product stewardship activities for recycling. Indeed, the early history of eco-labels addresses the biodegradability of detergent and soaps, which is a characteristically downstream (post-consumption) feature of the product.

There is, obviously, a close relationship between the corporate environmental reputation and the reputation of a specific product. Especially when there is a problem with a product, which will eventually generate bad reputation, it is almost impossible for the company to detach itself from it. Consumers and general public also associate highly visible brands like *Pepsi, Adidas or 3M* to its owner. But while this is true for the leading product of these companies, such as *Pepsi or Coke*, in order to restrain the effects of eventual product failure, specific brands and sub-brands are normally used to distance them from corporate brand name. Both Coca-Cola and Pepsi, for instance, own a multitude of soft drink brands that most consumers are not aware. Furthermore, only a few products of a large product portfolio managed by corporations are prone to present leading environmental prerogatives. Even fewer will hold an eco-label and are prone to be eco-branded. Hence, the distinction between process-oriented (or corporate-wide) standards and product-oriented standards, is helpful to uncover these kinds of intricacies involved in strategic environmental management.

## Eco-labels as vehicles for differentiation

The most elementary way of differentiating a product on the basis of its environmental prerogatives is to provide customers with information about it. The packaging can contain claims such as natural, ozone friendly, free-range, biodegradable or recyclable. This kind of information is certainly cheap to deploy, but there is a drawback. Self-certification or self-claim labels are certified only by the manufacturer; the reason the International Organization for Standardization (ISO)[10] does not consider them eco-labels but Type II labels. In other words, self-claims presented in Type II labels simply lack legitimacy. And, when it comes to environmental differentiation, reliable, uncontroversial information is crucial. At least for the *true blue greens*, who practice what they preach,[11] these types of self-proclaimed declarations are met with suspicion.

In contrast to the self-styled environmental symbol or statement made by a manufacturer or service provider, green consumers have more trust on a certified eco-label: "a label that identifies overall environmental preference of a product or service within a product category based on life cycle considerations".[12] Also important is the impartial third party accreditation of eco-labels to products that meet established environmental leadership criteria. In essence, eco-labeling programs aim to inform consumers which manufactures' products meet certain criteria for environmental impacts during the various stages of a product's life, from original manufacture to final disposal. Put simply, labels facilitate consumer choice.

The International Organization for Standardization considers only Type I (ISO 14020) and Type III (ISO 14025)[13] as eco-labels. Type I labeling is based on multi-party criteria agreement, which needs to be certified by a third party organization. Industry and consumers, for instance, agree on a set of criteria to judge the environmental performance of products and services during a transparent consultative process. The process is central for the credibility of the label. Type III goes further by requiring a Life Cycle Assessment (LCA) of the certified product by an independent third party. Since LCA generates quantitative data associated with the various impacts during production, use and post consumption, Type III labels work in a similar fashion nutrition labels do, which are also based on measurable variables, such as calories per grams.

Reflecting some pioneering environmentalism in Europe, the first eco-label developed was the German Blue Angel[14] in the late 1970s. Many countries followed suit and developed their own national eco-labels but most initiatives of the past decades such as the EU flower, the Australian Environmental Choice, the Nordic Swan and the FSC were multinationals and multi-stakeholders. The fundamental logic behind eco-labeled products is to differentiate them on the basis of above-average environmental performance. Eco-labels such as Blue Angel and the Nordic Swan should provide customers with sufficient credible information for their purchasing decisions. In many respects, eco-labels are expected to work in a similar fashion brands do. For example, the classic logo of the Italian *San Pellegrino* bottled water is sufficient for most consumers acquainted with it to ignore the rest of the information contained in the label. *Coca Cola*, as the most valuable brand in the world, *Sony*, *Apple* or *Puma* are also brands that tend to speak for themselves. Consumers familiar with these brands see the intrinsic product attributes and their emotional appeal, which do not require further interpretation.

If eco-labels could reach the same level of compact, intrinsic significance of normal commercial brands, its chances to succeed would be high. But there are some intricacies in eco-labels that distinguish them from commercial brands. Luxury brands such as the French *Louis Vuitton* normally try to relate products to their intrinsic and perceived qualities, aesthetics and emotional appeal. In other words, the brand appeals mostly to private benefits. Although eco-labels try to do the same, most of their intrinsic value relates to public benefits – of sustainable forestry management and biodiversity preservation, for instance. Hence, eco-labels transport the product (and the consumer mind) to realms much beyond their physicality. This is a journey that requires guidelines (standards) and, most importantly, trust. In some respects, while consumers may *stop* thinking when they see the *Louis Vuitton* brand – because there is tacit knowledge about its meaning – they may *start* thinking about the meaning of the purchase when confronted with an eco-label. In other words, eco-labels establish a relationship between private and public benefits, which consumers must believe as real, or at least worth the price they are asked for. Credible information is fundamental to establish this relationship and bring trust to the label. The next section explains why most eco-labels have not fully solved this problem.

## Credible information and eco-labels: Life Cycle Assessment

Type III labels (according to the ISO 14025) are the most demanding of the eco-labeling schemes. They require a reliable LCA: "a method for quantifying the environmental impact of an industrial process, activity or product".[15] The analysis comprehends the physical life cycle impact of activities from the extraction of raw materials or pre-manufacturing, manufacturing, use and final disposal of a product or the recycling of its component materials. Hence, it would be practically impossible to develop a reliable, measurable eco-label without the support of a LCA data, methods and tools. Hence credible eco-labeling depend very much on LCAs, which are supposed to provide reliable information. But LCAs carry their own problems.

The concept of LCA goes back to the early efforts at energy analysis in the 1960s, but the complexities involved in the design of an LCA tool resulted in a very slow process of acceptance by business and government. Distinct methodological assumptions are the main causes for the controversies between LCAs. Different definitions of the system boundaries of a product or service can cause substantial distortions in the results. Accounting for the total environmental impact of a product as simple as a pencil can become a very complex task. If a precise account of most environmental impact associated with the pencil is required - because it is practically impossible to consider them all - the scope of the LCA can be overwhelming. An all-inclusive accounting could include the air emissions of the chainsaws used to extract the wood that will eventually become a raw material for the pencil. Similarly, if a restaurant limits the boundaries of an LCA to the dining room and in-house kitchen, it can be expected that the environmental impact of its activities will be lower than a similar restaurant that extends the boundaries to the production of food and disposal of waste. Obviously, a sense of pragmatism suggests the demarcation of boundaries for such analysis, so that empirical results can be achieved. By establishing boundaries between systems, LCA become practical. Comparability between systems requires the use of similar scope and methodologies. However, this definition of scope is not free of political interests. The outcome of the LCA is the result of the inputs, and the inputs very often are a result of the preference of those who are paying for the study.

Life Cycle Assessment tends to emphasize the relative hazard that products cause to the environment. For instance, in a classical study developed in the 1970s[16] it was shown that, for the same number of bottles produced, Polyvinylchloride (PVC) consumes much less material and energy than glass bottles; they are responsible for less atmospheric and water-borne emissions, generating less solid waste than the glass bottles they replaced. The dumping of PVC in landfills, in this perspective, can be justified for its relative lower damage for the environment - in comparison to other alternatives. Hence LCA professionals tend to look for the best available solutions within current systems of production.

The availability of information also limits the diffusion of LCAs. Assessing the eco-performance of products requires data across their life cycle, which may reside in many separate sectors of the economy. The costs associated with data collection deter individual firms from generating all the necessary information. Besides, information may be commercially sensitive, unreliable or in an unusable form. This factor became clear in the development of the EU Eco-label project.[17] The development and acceptance of LCA has been crucial to the evolution of the EU Eco-label as a basis for the comparison of the quantitative environmental impacts of different commodities. Although there was consensus about the compilation of inventories, disagreement about methodologies for comparing the impacts of different emissions were always present. Accommodating such differences required a good degree of compromise among participants and, for truly green products, the resulting label may not be demanding enough. Such disputes and concerns are also present in a new type of eco-label.

## The emergence of carbon labels

Discussions about carbon labeling became more frequent in the second half of the 2000s.[18] In 2007 Sir Terry Leahy, the CEO of Tesco, a large United Kingdom (UK) supermarket retailer, announced that he was looking into labeling everything from food products to plants and flat screens sold in Tesco stores.[19] The talk got concrete in May 2008 when Tesco launched its carbon label initiative, with trial for 30 products.[20] The label was developed in collaboration with Carbon

Trust, an organization set up in 2001 by the UK government with the mission to search for solutions for a low-carbon economy.[21] Carbon Trust has been instrumental in developing and testing the first carbon-labels, which emerged out of collaboration with Walkers (potato-chips), Innocent (smoothies) and Boots (shampoo). Several other firms have subsequently joined the program; among them were Coca-Cola, British Sugars and Danone Waters. The ecological discourse partially explains such pro-activism, but the threat of future regulation on carbon emissions is an equally compelling rationale. In the case of Tesco, requiring suppliers to label their products is not such a risky move, since suppliers also have to bare the costs.

The initial candidates of carbon labeling were food products. The concept of Food Miles (FM) originated in the UK as a response to the increasing number of miles that food travel to get to the final consumer and the impact that it has on climate change.[22] The main factors behind the increase in miles transported are free trade agreements, new consumer habits, geographical centralization of food production and transportation costs. The main strength of the FM concept is the easy access to data, so the effect on the environment due to carbon emissions can be calculated with the Weighted Average Source Distance (WASD).[23] The FM concept is a relatively simple and user-friendly label that helps consumers to make informed choices about the embedded carbon emissions in product due to transportation. However, few miles do not necessarily mean few carbons. For example, the total carbon emissions of heating up greenhouses to grow roses in the Netherlands are higher than growing them in Kenya and flying them to Amsterdam. The FM concept has also been criticized not only for ignoring carbon emissions during production, but also for the potential of being a veiled argument for protectionism against developing countries.

The Carbon Footprint (CF) concept addresses the weaknesses of FM by using LCAs to calculate the total amount of carbon emissions throughout the entire life cycle of products. However, as it was mentioned earlier, it is difficult to calculate the CF with accuracy because primary data for different stages of a LCA may simply not be available. Indeed, the methodological issues of LCAs are certainly present in carbon labels. For instance, the largest part of carbon emissions may occur while transporting the product to the home of consumers and/or preparing and conserving it. Should the LCA include these stages, which are not under the control of producers? Such issues explain the heated debate about LCA methodologies, and the difficulty of developing standards for CF.

There are other concerns about the efficacy of climate labels. Focusing on carbon emissions may divert consumer's attention from other negative environmental impacts. By their very nature, carbon labels narrow down the focus to a single problem, disregarding – or at least downplaying – other impacts on land and water, including biodiversity loss and pollution. Low levels of carbon emissions may be a result of the use of nuclear power, masking problems associated with accidents and waste disposal. Low-carbon emission products might contain chemicals that are dangerous to the environment and consumers' health. Finally, concerns have also been raised about the types of Greenhouse Gasses (GHG) emissions that should be included in climate labels. Besides Carbon Dioxide ($CO_2$), food production (intensive agriculture, in particular) entails the emission of other potent GHG such as Methane ($CH_4$), Sulfur Hexafluoride ($SF_6$), Hydrofluorcarbons (HFCs), Perfluorcarbons (PFCs) and Nitrous Oxide ($N_2O$), the last one being 310 times more potent than $CO_2$. An ideal climate label should consider all these GHG during the entire life cycle of the products. However, as we saw, methodological hurdles may limit such goal, as a Swedish initiative suggests. The Swedish National Labeling Institute, the Swedish eco-label for organic food (KRAV) and *Svenskt Sigill*, a label created by the Swedish farmers' association, have been working on the development and testing of climate friendly and carbon neutral labels.[24] As one could expect, the initiative has not been friction-free. There is no agreement, for instance, on what carbon neutrality actually means. Some experts even argue that there is no such thing as carbon neutrality.[25] As a result, the vast majority of producers will keep waiting until a standard for climate label emerges.

Overall, the future of eco-labels depends very much on practicalities. At present, they seem to cause confusion both to retailers and consumers. Retailers face the problem of finding decent ways of calculating and measuring progress and communicating it to end-consumers in a trustful way. This is especially true for carbon labels. Consumers, on the other hand, are uncertain about what labels stand for, and the fear of being fooled into paying a premium price for something that they are not able to appreciate properly. Most of us will not know whether a label showing 400g of $CO_2$ in a potato chip package is good or bad, even if we have other products to compare with. These controversies and trade-offs suggest the need to incorporate climate labels into existing eco-labels, so that the consumer could make a choice while taking the complete environmental impact of the product into account. However, integrating fair trade and climate calculations in the same label also requires collaboration among different labeling schemes and eventually some amalgamation among them. Not an easy call.

## Eco-labeling success: to whom?

Originally, eco-labels have been thought of as having potential benefits for consumers as a way of limiting confusing, extravagant or false claims of product environmental performance, acting as leverage to foster industry rivalry and reduce industrial impact on the environment. Firms would compete on the basis of eco-attributes of products, systematically raising the bar, which would lead to more sustainable industries and societies. In practice, eco-labeling schemes had only a partial success. Apart from relative success in Scandinavia and Germany, most European eco-labeling projects face difficulties in achieving their main purpose of environmental marketing differentiation. Overall, from the immense array of products currently marketed, only a very small percentage adopts eco-labels.

Apparently, there is no direct relationship between certification to an eco-label and shopping behavior. When asked, most consumers describe themselves as conscious consumers who prefer eco-friendly and ethical products. In practice, however, only a minority convert their assertion into purchase.[26] A fragmented consumer response to ecological appeals is typical, explaining the reluctance of many companies in applying to eco-labeling programs. Since companies already pay employees to test products, the cost associated with certification by a third party organization is difficult to justify. Besides, European eco-labeling schemes also face red tape. It takes on average more than a year for some products to be certified – longer than the marketing life cycle of many goods.

Similar to the Green Clubs mentioned in Chapter 4, eco-labels seem to work better in avoiding disadvantages than generating competitive advantages. Suppliers of toilet paper in Scandinavia can tell it. Today it is practically impossible to enter that market without an eco-labeled product. In many other product categories, eco-labels became just a license to operate. The Forest Stewardship Council (FSC) label is also exemplary. As it was suggested in the case opening the chapter, FSC served better the interests of forestry companies, since the label guaranteed privileged access to their business customers, rather than serving retailers to reach final consumers. The history of the label partially explains this outcome. After all, FSC emerged as a response of forestry companies to strong eco-activist campaigns. Pressed by activists, large retailers such as Home Depot in the USA passed the responsibility (and costs) of supplying certified wood onto forestry companies. For the suppliers of the Swedish retailer IKEA, for instance, not having FSC-certified wood/products meant a clear disadvantage. For suppliers who obtained the certification, the eco-labeling scheme guaranteed their contracts, even if at reduced profits. In other words, even though the FSC label is used in final products, it represents an agreement between a civil society organization (the WWF), forestry companies and retailers. FSC standards and labeling represent a clear private regulation for wood and wood products in Europe and US that served mainly B2B or industrial markets transactions. Consumers were mostly absent of the negotiations resulting in the FSC label. Not surprisingly, they were not willing to pay more for FSC-certified products. The costs of compliance remained within the production chain and had to be shared among forestry companies and retailers.

Overall, when it comes to the success of eco-label schemes, the evaluation tends to be confusing. The reason is simple: the criterion used by the promoters of the labels is very different from its users (companies). For the promoters of the FSC or the Nordic Swan, for instance, success is measured by the number of products adopting it; a higher number of products with eco-labels (high diffusion) makes the scheme more successful. On the other hand, companies that use the eco-labels in their products for the purposes of eco-differentiation are focusing on uniqueness. A high adoption rate of eco-labels by competing products erodes the exclusivity and eventual competitive advantage brought by the eco-label. For companies, the lower the diffusion of eco-labels, the better the chances of eco-differentiation to pay off. Therefore, there is an intrinsic contradiction in the interests of the promoters of the schemes and their users. As Reinhardt[27] has conveniently emphasized:

> By putting pressure on other firms in the industry to adopt environmentally preferable production of product characteristics, eco-labels may hasten imitation and erode the differentiating firm's market position. A firm seeking to differentiate products along environmental lines needs to consider carefully whether its competitive position will be improved or impaired by eco-labels.

The lesson here is clear: As eco-label schemes become more successful, the competitive advantage of user products will gradually be dissipated.

Although eco-labels will remain instrumental in eco-branding strategies, they may not be sufficient to generate competitive advantage. Even when they do, companies should not rely on them for long-term differentiation. For that, it is necessary to reconsider what eco-labels are supposed to do at the first place – to inform consumers that products or services are unique also because they do the right thing. As the following cases suggest, environmental issues may increasingly become part of the brand contract between the consumer and the company.[28]

## FROM ECO-LABELING TO ECO-BRANDING

In Sweden, consumer environmental awareness is remarkably high.[29] Eco-labeled products are literally ubiquitous and differentiating on the basis of ecological prerogatives is becoming increasingly difficult and costly. In this context, one of the largest retailers of food and domestic products developed a creative way of differentiating a portfolio of products. Coop Sverige, the owner of 355 city supermarkets with a clear orientation toward ecological excellence, and 40 more generalist hypermarkets created *Änglamark*, an ecological brand to communicate the image of environmental responsibility of more than 250 food and domestic products, such as coffee filters and washing powder, sold by both Konsum and Forum. *Änglamark* sales increased from €3 million in 1991 to a yearly average of €55 million between 2004 and 2008, conferring the leadership of the eco-brand in the Swedish market.

Food retailing in Sweden is concentrated in three large firms: ICA, Coop and Axfood. Together, these companies control more than 95 percent of the market. ICA is the dominant player with almost 50 percent of the market whereas Coop is the second largest player with approximately 23 percent. Although there are only a few players, the market is ruled by fierce price rivalry. The trade-off between price and environmental attributes of products is a constant challenge for consumer retention. Lidl, a German retailer that entered the Scandinavian market in late 2003, embodies this challenge. An aggressive strategy resulted in Lidl opening stores in a few dozen Swedish cities, with several more stores planned to open in the coming years. The German retailer is very competitive in the low-price segments, benefiting from economies of scale to supply its stores with products shipped from a central warehouse in Germany.

In developing the *Änglamark*, Coop elaborated the values and the functions that ecological products should have in order to succeed in the marketplace. Although the brand name *Änglamark* evoke positive values associated with the Swedish soul, identity and with nature, therefore appealing well to the greenies, Coop also had a very commercial sense of the brand. Eco-products should taste as good and perform as well as conventional products but should not necessarily be more expensive, and customers should easily find them in the stores. The product mix included both food and non-food products and was developed according to clear criteria. Products should fulfill the requirements of eco-label schemes such as the Nordic Swan, the EU Flower or KRAV (the Swedish eco-label for organic food). In other words, most *Änglamark* products should be endorsed by eco-labels; present the lowest possible $CO_2$ footprint and minimal packaging, preferentially made with renewable material; be produced in accordance with ethical values; contribute to higher levels of well-being and must be equal to the product leader in terms of quality.

In order to sell high volumes of eco-products, Coop used *Änglamark* to compete directly with market leaders in each product category. In its first five years (1991–1996), *Änglamark* was the first organic alternative in all food and often non-food categories, covering more than 80 percent of the product range in the supermarkets. Such position gave Coop a first-mover advantage in strategic product categories – those that, eventually, became an issue for the media and customers. Detergents were such a category at the beginning of the 1990s, requiring manufacturers to develop and market biodegradable and eco-labeled products. Since *Änglamark* offered products that performed as well as the conventional ones, it gradually gained consumers preference. As a result, the meaning associated with the name, the intrinsic characteristics of the products and the strategy for product mix and pricing guarantee an initial success for the eco-brand.

One of the keys to success of the eco-brand – but also a main challenge – was finding, developing and managing reliable suppliers. For instance, at the beginning of the 1990s, there were few suppliers of organic food. Most of them were small and specialized pioneers and some were highly dependent on Coop. Many of them helped Coop to develop the competences and the assortment of eco-products. This close collaboration made it easy for the retailer to influence the production processes and product safety, which required risk management plans according to the Hazardous Critical Control Points (HCCP) at an early stage. Since the whole chain has to be certified – all the way back to the producer of raw materials – ecological certification of suppliers is complex and time consuming. Hence, the close collaboration between Coop and its suppliers reduced the costs of compliance when HCCP became regulation, representing a first-mover advantage for some.

By 2008 *Änglamark* was still Coop's central ecological brand. The total range of *Änglamark* has remained at around 200 products but in some product categories, Coop has withdrawn it because some producers developed their own eco-brands. Coop continues to reinforce its sustainability image with biodegradable plastic bags for shoppers and making easier for customers to audit how much they have spent in eco-products. *Änglamark* certainly helps Coop to maintain its green credentials in Scandinavia but the late entrant Lidl provoked some price wars. Rivalry in certain product categories such as milk, coffee, mineral water, flour, sugar, salt, frozen meals, detergents, fish and meat are on the rise. Will Scandinavian consumers remain committed to green consumption or will price wars determine the future of eco-branding?

## Building an eco-brand[30]

On the other side of the globe, a different kind of business has been working on eco-branding strategies. Lend Lease Australia is a large company that focuses on the development, construction and management of real estate assets for both public and private enterprises. In the early 2000s Lend Lease was going through difficult times. The public image of the company was taking a battering in the press; after a poorly conceived attempt to enter the US property market, its share price was at around A$10 (€5.2) from a high of around A$23 (€12). It was time to rethink the corporate strategy and rebuild shareholder confidence.

During this period, Lend Lease initiated the review process that lead to the construction of the *Bond*, as the new headquarters of the company is known in Australia, concluded at the end of 2002. The senior management decided to use the new building to support a strategy that should revitalize Lend Lease and re-establish its status as the leading property group in the industry. The building was to be iconic and different. It was to demonstrate Lend Lease at its best, reflect its core values of respect, integrity, innovation, collaboration and excellence and showcase the future of building design. It was this latter requirement that brought sustainability into the design process, also identified as an emerging market opportunity. Designing a sustainable building was seen as an opportunity for Lend Lease to exercise industry leadership once more and to differentiate itself from competitors by showcasing their new product – the *Bond* – as their headquarters.

The headquarters was designed by Lend Lease to be a healthy building as well as to meet the strictest requirements and demonstrate ecologically sustainable principles, resulting in a building with several key eco-friendly attributes. By designing an energy efficient building, the architects also provided a socially environment for the occupants and the community. Eco-friendly materials, processes and design features were extensively used: double-layered insulated glass coupled with external blinds to reduce cooling costs; electrical design that used an open-plan lighting system to reduce the consumption of electricity and support the use of natural light; mineral-based interior paint that is non-toxic and solvent free; bamboo floors produced with water-based zero-emission coatings; non-toxic glues and waste-free manufacturing processes, perforated ceilings were incorporated into the design to accommodate the chilled beam air-conditioning system and the ceiling tiles were manufactured from steel with a 25 percent recycled content. All included, the building has one of the lowest running and maintenance costs of any building of its size. It was the first office building in Australia to receive a 5-star greenhouse rating under the Australian Greenhouse Buildings Ratings Scheme.[31] The *Bond* is commercial in style and aesthetics and yet is a green building. This has helped overcoming the perception that green buildings simply are not suitable for commercial use, or that they somehow have to look odd.

Overall, the *Bond* was a remarkable marketing success. By 2006, thirty four awards were conferred to the building. In terms of media coverage, the investment largely exceeded the expectations. The attention received by the media resulted in free marketing, increasing the value of the Lend Lease brand and influenced its share prices. Another surprising intangible outcome comprised happier employees and increased productivity levels. In an internal survey, 85 percent of the staff said that they work more comfortably and more than half asserted that they are more productive in the new building. Such indicators appear also to have influenced employee retention.

Curiously, despite the success of the *Bond*, Lend Lease faced a challenge in the further commercialization of green buildings. The diffused ownership structure that characterizes the Australian commercial property sector poses some limitations on the ambitions of Lend Lease. The ownership structure can make it very difficult to sell ecological architecture. This is because most buildings are not owned by the occupants but rather by large investment funds. The majority adopt the net lease system (also used by the *Bond*) in which the customer pays for the costs of running the building (electricity, water and other maintenance costs). In this system, reductions on the costs of running the building benefit the tenants, not the owner. The gross lease is an alternative to this system. In this case, because the running costs are included in the rent (hence paid by the owner but transferred to the tenet via a gross bill), the landlord has also interest in reducing operational costs during tenancy. Such system, however, is not as common as Lend Lease wishes. Although the *Bond* certainly helped Lend Lease to establish an eco-branded product, expanding the market for green buildings depends on eventual changes in regulatory frameworks, consumer behavior, as well as the contacting system of real estate markets.

## WHEN *ECO-BRANDING* PAYS

The two cases suggest that eco-branding is hard work. According to Joel Makeover: "Given a choice, most consumers will be happy to choose the greener product – provided it does not cost any more, comes from a trusted maker, requires no special effort to buy or use and is at least as good as the alternative. That is a hard hurdle for any product".[32] Eco-branding implies distinctiveness, an attribute that, by its very nature, is only achieved by a minority.

As in the other Competitive Environmental Strategies (CES) discussed in the preceding chapters, the conditions that satisfy eco-branding depend on variables ranging from the structure of the industry, the regulatory framework and the capabilities of the firm. These generic conditions provide the broad context in which a corporation might decide to explore this strategy. However, when compared with the other CES, in the case of Strategy 3 the context can be more clearly defined. Forest Reinhardt[33] has previously identified three requirements for environmental product differentiation. Eco-branding strategies have the potential to generate competitive advantage when: reliable uncontroversial information about product's environmental performance is available to the consumer; the differentiation is difficult to be imitated by competitors, and; consumers are willing to pay for the costs of ecological differentiation. The following sections revisit these requirements in the light of the discussion about eco-labeling schemes and the cases presented in the chapter. An additional aspect facilitating eco-branding complements the three requirements suggested by Reinhardt: the convergence between public and private benefits.


## Context

Credible information is the basic prerequisite for environmental product differentiation. In the case about eco-branding in Sweden, the owner of the brand, Coop, extensively uses the eco-label KRAV[34] to endorse organically grown food products sold in the supermarkets. Coop charges between 10 and 100 percent higher prices than similar products that are not certified by KRAV. The fact that KRAV is accredited by the International Federation of Organic Agriculture Movement and controlled by the Swedish Board of Agriculture confers a high degree of credibility to products with this label. This was the rationale for *Änglamark* food products to use the KRAV label; it confers credibility to the environmental claims via third party certification.[35] Besides KRAV, Coop also uses other eco-labels, such as the EU Flower, the Scandinavian Swan and the Fair Trade, resulting in some *Änglamark* products presenting three or more eco-labels.

Third party certified eco-labels certainly confer legitimacy to the claims of *Änglamark* products. They simplify the data and reduce the complexities involved in informing the eco-attributes of the food and non-food products. But even if eco-labels are crucial elements to provide legitimacy for *Änglamark* products, they are not sufficient for the success of Coop's eco-branding strategy. As in other commercial brands, substantial marketing efforts are necessary to promote eco-brands so the imagery of its logo can become synonymous of environmental responsibility. For instance, rather than the eco-labels, the most visible element in all *Änglamark* products is its "Ä" logo, suggesting consumers to trust the eco-branded products as the leaders in environmental performance. Overall, credible information is brought by certified eco-labels but building an eco-brand requires more than that: overtime, it is necessary to build trust in the brand in a way that the importance of eco-labels diminish. Ultimately, what companies want from their consumer is trust, no matter how much information they provide.

The Lend Lease case makes this point clearer. The *Bond* was the first office building in Australia to receive a five stars greenhouse label by the Australian Greenhouse Buildings Ratings Scheme. However, in that case, the eco-label had only a minor importance in the success of the Eco-branding strategy of Lend Lease. The product (building) drew attention to itself in several fronts. Besides the extensive consultation process taken during the design and construction phases, several innovative technologies used in the building, such as the chilled beam air-conditioning system, were brought to the attention of media and general public. After receiving more than 30 awards, the building became iconic. The *Bond* represented the cutting edge of ecological architecture and an example of how office building should look like in the 21st century. Even though the concept of green buildings is not particularly new, Lend Lease managed to associate ecological responsibility to its brand by building trust. Starting with employees and neighbors, who we consulted about the project, the claims of technical efficiency and performance were genuine and the most advanced in the country at that time. In that case, credible information became trust, literally.

Barriers to Imitation is the second prerequisite for eco-branding. Limiting competitors from copying (the attributes of) products is also central for the success of eco-branding strategies. If product environmental differentiation is to be successful, eco-innovation should not be easily replicated. In markets in which products do not present good environmental performance, an eco-label may be sufficient to keep competitors at bay at least for a while. After all, for the newcomers, complying with the prerequisites of an eco-label requires substantial resources and time, so the product leader should be able to differentiate just by holding an eco-label. However, as it was mentioned earlier, such differentiation will erode as more products acquire the same label. Scandinavia, as one of the most advanced areas in the world in green consumption is exemplary in this respect. In sectors such as household cleaning products and toilet paper, the diffusion of eco-labels is so high that rather than being differentiators, eco-labels just guarantee a license to operate.

Once again, the clear lesson in this case is that barriers to imitation require a protective patent of one form or another.[36] For industrial products, holding an exclusive rights certificate will reduce the chances of imitation. This is obviously neither new nor exclusive to eco-oriented products. Hence, the logic here is the same for traditional products. On the other hand, when it comes to consumer products, consumables in particular, the long-term differentiation strategy may require the development of a commercial brand. As the marketing strategy adopted by Coop in Sweden suggested, barriers to imitation were brought by the brand, rather than by the eco-friendliness of the products *per se*. Although rival supermarkets can also sell eco-labeled products, imitating the eco-brand is practically impossible – at least not without substantial resources and time.

Willingness to pay is the third prerequisite for eco-differentiation. Consumers need to perceive a clear benefit for their purchase. In the case of industrial markets, the benefits are normally translated into cost savings, better performance of the product (as an input for another industrial process) and cost reduction of risk management.[37] For instance, equipment and machinery that consume less energy and reprocess by-products might reduce the costs of operation for the client. As General Electric (GE) has done so well with the *ecomagination* program, the vendor company can explore both the ecological (public benefits) and efficiency-related (private profits) attributes of products, which may result in advantages during use. As a result, price premiums can be obtained in some cases (think of a press shop machine that consumes less energy). However, the total savings the product will be able to make for the buyer/user frequently limits the magnitude of price premiums. In the B2B case, there are higher levels of rationality guiding both the evaluation of the product and the willingness to pay higher prices. On the other hand, in consumer markets, the attributes associated with the products may result in relatively lower private benefits for the consumer (think of a biodegradable cleaning product), hence requiring companies to make substantial marketing efforts to link the image of ecological responsibility via an eco-brand. For both, industrial and consumer markets, however, it is essential that the consumer is willing to pay for ecological differentiation.

Ideally, the three requirements for environmental product differentiation should be complemented by another element: the convergence between public and private benefits. Although eco-branding strategies can succeed without this convergence, it certainly increases the chances of success. The reason is simple: converging benefits eliminates some of the trade-offs involved in the purchase of eco-oriented products. Observe the case of organic food. Although most people may say that they buy organic food because is good for the environment (public benefit), they also know that they are buying healthier food, which represents a private benefit. In this case the classic dissonance between what people say about their willingness to pay for eco-products and their actual shopping behavior is minimized because there is no trade-off between paying more to protect the environment (via organic food) or paying less for non-organic food. What most consumers are buying, really, is the private benefit of healthier tomatoes, bananas or grapefruit. Coincidentally in this case, the public benefits of sustainable horticulture also results in healthy fruits and vegetables. This is an undisputable win–win scenario, which partially explains the exponential growth of the organic food segment in the past decade or so, which also facilitates the lives of marketers.

That is good news for those in the organic food business but most initiatives serve the interests of environmentalists and business very differently. As it has been addressed earlier, the diffusion of eco-labels makes a classic case. The rationale for governments and eco-activist groups to establish an eco-label scheme relates to the possibility of extending the public benefits to as many products and companies as possible. They take, fundamentally, the environmentalist's perspective. On the other hand, companies applying for eco-labels are trying to differentiate their products, envisioning the price premiums that eco-differentiation might bring. Unfortunately, for both environmentalists and business, the convergence between private and public benefits is extremely hard to be artificially created. As the organic food example suggest, this is more a coincidence than an outcome of managerial ingenuity.

## Competences

Broadly, developing eco-branding strategies require companies to have the traditional competences in brand management, as well as a good understanding of the main issues associated with green marketing behavior. Although the traditional knowledge in marketing is certainly a good point of departure, success of failure of Eco-branding strategies are often in the details. In this case, the details may be hidden in the answers for the following questions:

- Do we have enough knowledge about the environmental impacts of our products or services? Does anyone in the company have enough knowledge about Life Cycle Assessment (LCA) or eco-labeling?
- Considering the characteristics of our products, can we differentiate some of them on the basis of ecological prerogatives? What would be the rationale behind such differentiation?
- Is our knowledge about traditional branding sufficient to develop an eco-oriented brand? If not, how can we develop it?
- What would be necessary to develop an eco-brand? Do the products already present the necessary requirements? How much would be necessary to bring them to the level of green products?

- In the case of developing an eco-differentiation, can our competitors imitate it with relatively few resources and capabilities? Why?
- Can the endorsement of an eco-label represent a differentiator for our products? What about the rival products; do they present eco-labels? How difficult would it be for competitors to obtain an eco-label?
- Do we know the carbon footprint of our products? If climate labels were required from us, what would be the situation of our products compared to rival products? Does anyone in the company have enough knowledge about methodologies for carbon labeling?
- Does the eco-differentiation of any of our products depend on sophisticated and controversial information? Can we by-pass such problems by working in the upstream or downstream value chain? Do we need to invest substantial resources to achieve such goal?
- Can the company obtain price premiums for eco-differentiation? Why would consumers be willing to pay for that? How do we know it? Is the knowledge we have about our customers sufficient to know their environmental profile?

## CONCLUSION

The consumer has often been identified as the major engine for the change toward sustainable societies. In the act of shopping, they vote pro or against companies they consider good or bad citizens. When expressing their intentions to reward green companies in surveys, consumers have been consistent. They said over and again they would privilege greener products and eventually pay more for them. Many even said that, when offered the choice, they would prefer more free time rather than more money for working longer.

Empirical evidence undermines such claims.[38] Studies about material intensity and affluence show that increases of consumption levels have a direct impact on resources and, indirectly, on the environment. Considering that systematic increasing consumption levels are not ecologically sustainable, there were hopes that, as we become richer, we would prefer to buy time instead of physical goods. In reality, increases in productivity levels have historically been transformed into increases in income, instead of leisure. As people become more affluent, they tend to buy more goods and services with higher material intensity, rather than more eco-friendly ones. In other words, affluence and resource intensity work in tandem. A clear example is the increases in food prices in early 2008 as a result of the affluence of middle classes in emerging economies, China and India in particular.[39] Affluence allows people to eat more meat, and (grain-fed) meat production requires disproportionately higher productions of cereal. The result is higher food prices.

There are three sets of explanations for the increasingly higher resource intensity behaviors.[40] The first comprises the economic and socio-economic aspects of the institutional set up of the economy. The high levels of consumption of industrial products relates to the persistent fall in their prices, compared with products and services that cannot be provided industrially. Prices of electronic gadgets such as DVDs and cell phones decreased substantially over time, and people with lower incomes could eventually afford them. The second explanation focus on consumption from the socio-psychological perspective, in which human beings are embedded in particular social relations. Consumerism can be driven by variables ranging from envy to the need for people to make sense of their own lives and justify a personal self-image:[41] "As goods are used as markers and classifiers, they make visible and stabilize the categories of culture – they, so to say, constitute the visible part of the culture as the tip of the iceberg which is the whole of the social processes". The third set of explanations comprises historical and socio-technological elements of different aspects of everyday life. Although consumption is often discussed as a manner of choice, socio-technical embedding actually bound most choices. Overall, shopping habits are embedded in both inside consumer's psychology and in the external organization of society.

Eco-oriented consumption is just one part of the complexities of economic, social and political interpellations constituting consuming subjects. While citizens may seek to relate practice to ideology, their political organization, lifestyle and consumption decisions always display tension. We shop and consume more typically as complex and embedded social individuals rather than as coherent and fully ideologically formed members of a party or a social movement. As consumers, we do not find it difficult to recognize that it is not easy being green. The normal conditions of existence for most of us involve uncertainty, equivocally and a cacophony of competing interpellations. In part, that is the attraction of market pluralism. Many disparate signs jostle for our attention. The market is overwhelmingly an economy of significations that seeks to enlist subjectivities: as fashionable, as dynamic, as caring, as feminine, as masculine and, eventually, as green. For most of us, there is too much information to consider, too many problems to bother with and too many self contradictions in the understanding of what is the right thing to do.

The lack of objective economic constraints or rewards in regard to the purchase of green products also seems to reinforce their most direct convenience for the consumer. The intrinsic benefits of most goods and services address individual self-interests, rather than environmental responsibility. As a result, environmental solidarity, as a socially preferred condition (public benefits), is still absent from the shopping list of most consumers. What people say about their willingness to buy green products normally differs substantially from what they do in their actual purchasing behavior. "Individuals have incompatible beliefs, and do not rank them in a single hierarchy in the same manner of the rational man of economic theory. Citizen preferences are judgments about what *we* should do, while consumer preferences are expressions of what *I* want".[42]

Nonetheless, we cannot deny the potential a specific set of consumers represent for successful eco-branding strategies. As this chapter explored in detail, there is an undeniable space for incorporating ecological concerns into product design. Such power, however, has to be put into a grounded business perspective. Pro-environment behaviors have often been more an ideological hope than to the reality expressed in the shopping habits of mass consumption.[43] Green consumerism remains a niche market mainly because of the complexities associated with the wide range of factors influencing decision-making. The successful ecological differentiation of products and services demand the alignment of at least some of these factors. The mechanisms of political organization that are essential to its success are more easily available to firms than to disorganized consumerism. Consumers rely on the ability of corporations to present compelling alternatives to the current range of products. This is good news for companies that understand the intricacies involved in eco-labeling, as well as the elements required for the successful deployment of Eco-branding strategies. As some examples discussed in this chapter suggest, intangible aspects such as symbolism and trust are central in establishing a lasting relationship between eco-branded products and consumers.

# 6

# ENVIRONMENTAL COST LEADERSHIP

Companies strive to distinguish their products and services. They do their best to present them with features that consumers eventually value higher than rivals. Such efforts often lead to higher costs, and aiming at price premiums via differentiation strategies is a viable solution for these companies to cover such costs. When it comes to eco-oriented products, it is not much different. As Chapter 5 explored in detail, if being green costs more, the company has little choice but trying to obtain returns from eco-investments via Eco-branding strategies. This is okay for firms that are able to tap into niches for eco-oriented products but, by their very nature, niches represent only a small slice of the market. No matter the efforts companies make, markets have limited scope for differentiation. Industrial markets (or business-to-business – B2B), in particular, have a very strict sense of costing, and obtaining price premiums is normally attached to eventual savings during product use. In other words, no matter how eco-friendly a product is, when competing in price-sensitive markets, it has to be cheap first.

Does this mean that products and services that need to compete on the basis of low cost will never be able to offset eco-investments? Can firms overcome the trade-offs between eco-investments and costs? Indeed, this has been the critical challenge encompassed in the *pays to be green* debate. A few companies have been competent enough to develop products and services that present both reduced cost and environmental impacts, but this is certainly a tough call for most. Unless companies are able to deploy innovative designs, alternative materials or even market their products in a different manner, eco-investments may result in higher costs, which would restrict the deployment of Environmental Cost Leadership strategies (or E-cost, for short). For a selected number of companies described in this chapter, the tough call of E-cost has been possible only after lots of efforts. Even so, they inspire others to pursue similar strategies. Since clients tend to privilege low costs, companies that are able to offer eco-attributes at low prices in their product portfolios are in a much better position to compete in tougher regulatory environments. The opening case is instructive in showing this aspect.

## ECO-DESIGNING PRODUCTS: LIFE CYCLE THINKING

Ecolean is a relatively young packaging manufacturer that grew extremely fast after starting its operation in 1997 in Helsingborg, in the southern tip of Sweden.[1] The company has representatives in 20 (mainly developing) countries, having grown an average of 50 percent per year since it was funded. The company sells around 250 million units of packaging per year, generating around 30 million US$ in revenues.[2] Ecolean's main products are filling systems and packaging films for its own designed stand-up pouches for liquid foodstuffs, which predominantly are sold to emerging economies. Ecolean supplies wrapping films for butter in France (Carrefour) and UK (M&S) and for sausages in UK (Tesco), form-fill-seal foils and films for dip sauces in portion packs for McDonald's outlets in UK, Scandinavia and Russia and flow-pack films for potato chips in Sweden.

In average, Ecolean packaging not only costs 25 percent less than competitors but also presents the lowest environmental impact. This is possible because the company adopted a radically new proposal for packaging. Between 40 and 60 percent of oil-based *plastics* (HDPE: High Density Polyethylene and PP: Polypropylene) used in packaging was substituted with calcium carbonate as raw material ($CaCO_3$ – most commonly known as chalk). Besides being one of the most abundant minerals in the earth crust,[3] calcium carbonate does not present any toxicity – the reason for the Federal Drug Administration (FDA) to classify it as Generally Recognized as Safe (GRAS) for human beings.

The environmental advantages of substituting polyolefins (HDPE and PP) with chalk are many. A Life-Cycle Assessment (LCA) concluded that the environmental impact of Ecolean products are substantially lower than competing materials (plastics, cartoon and aluminum) in all categories (water and energy use, emissions, etc.) during all phases of the product life-cycle.[4] Besides, the use of calcium carbonate results in Ecolean packaging being biodegradable under certain conditions (it needs to be exposed to light). But because today most solid waste in developing countries – the main market for Ecolean products – is landfilled or incinerated, the company does not make any claims in this respect. It neither makes claims about the additional benefits of Ecolean packaging when incinerated: the calcium carbonate reduces the acidity of fumes.

Since Ecolean is less damaging than other products, one may wonder why the company is not more aggressive in marketing the eco-attributes of its products. The answer is quite simple: because the clients would not pay for it. Although clients consider the environmental features of Ecolean products a good thing, the products have first and foremost to fulfill function requirements at a very competitive price. All satisfied, environmental features are a plus. Nonetheless, as the production of oil peaks, causing prices to remain high, as well as post-consumer waste regulations become more demanding, such strict view might change. Government representatives in China, for instance, where Ecolean installed a new factory in 2001, showed interest in the product for its lower dependence on petrochemicals and the abundance of calcium carbonate in Chinese soil. The environmental attributes of the packaging are slowly giving Ecolean a first-mover advantage. Of course, competitors can eventually copy this innovation in material substitution. But since Ecolean also manufactures and sells filling machines for its own stand-up pouch (which has been patented), it has acquired substantial competences in the use of the new raw material.

## Eco-design as a method and management tool

As the case suggests, in a great number of industries, competition tends to be based on price, leaving little room for suppliers to charge for eco-differentiation. In the packaging industry, regulatory measures such as post-consumer taxes have been on the rise in developed countries and in the coming decades are expected to be tighter even in emerging economies. This means that packaging material will have to be competitive on price and environmental performance. The combination of low margins with the saturation of mature markets in many industrialized countries increases rivalry and put manufacturers under extreme pressure to reduce costs. In addition, an increasingly demanding customer and a never-ending tightening of environmental regulations make competition even harder. For firms operating in such context, focusing on E-cost strategies may be the only choice to generate competitive advantages.

The case of Ecolean also leads us to the so-called concept of *eco-design*,[5] which in simple terms, means the design of products with the intention to reduce their embedded environmental impact (or load). For instance, the substitution of chalk ($CaCO_3$) for plastics could be considered an eco-design exercise at Ecolean, since the specification of the alternative material in the design phase, beside cost savings, had also the aim of reducing the environmental impact of the products. Eco-design is often helped by Life Cycle Assessment (LCA), discussed in Chapter 5, which is both a methodology and a tool for the identification of environmental impacts associated with the life cycle of a product. Designers often use LCAs to compare the environmental impacts of products in the same category but LCAs can also be used to legitimate the claims of eco-friendliness of products, as Ecolean does. Overall, LCAs are important tools for companies to apply eco-design principles and practices.[6]

The packaging industry is also a good example of one of the eco-design principles: dematerialization. The environmental impact of post-consumption has been the target of regulatory measurers, motivating designers to work toward the reduction and substitution of materials to facilitate reuse and recycling. As a result, many products that used Styrofoam for casing eliminated it by simply redesigning the cardboard boxes. Mobile phones and other electronic gadgets are among many electronic products that nowadays are transported in cheaper and more eco-friendly packaging. Clever design that reduces or eliminates non-recyclable material (Styrofoam, in the case) and facilitates recycling (of cardboard) has often the additional benefit of costing less. Hence, dematerialization has the obvious advantages of reducing both environmental load and costs of products. For companies operating in the packaging sector, competences in eco-design are then imperatives to deploy E-cost strategies.

Eco-design can also reduce the weight or volume of materials in a product. The Swedish retailer IKEA is well known for its flat packaging concept, which optimizes volume, hence reducing costs and environmental impacts during transport. Reduction of environmental impacts arising from the use and maintenance of the product is also a straightforward result of eco-design. The success of the General Electric (GE) *ecomagination* program, for instance, has much to do with the energy savings during the use phase of products supplied by GE. Another aim of eco-design is to design products that can be reused, remanufactured or recycled. A product made of eco-friendly material (non-toxic, renewable, reusable or recyclable parts), with reduced weight and lean packaging may result cheaper. In relatively simple products such as packaging, cleaning products and materials used in the finishing of buildings, the benefit of applying eco-design is often self-evident.

The case of the *Bond* building in Australia, presented in the previous chapter, is instructive in showing the advantages of eco-friendly materials. The paint used in the building was mineral-based and solvent free, the bamboo floors used zero-emission water-based coatings and non-toxic glues made a healthy building. The demand for healthy and energy-efficient buildings has been growing steadily in the recent years and, for the suppliers of non-toxic materials, the market is booming. For those who are able to develop eco-attributes at low costs, competitive advantages are prone to emerge. Would this be the case for more complex products such as computers, refrigerators or cars? Can mobile phones, music players and plasma screens compete on the basis of E-cost strategies? When does design for disassembling such products pay? When are the efforts to recover materials and recycle parts worth? The case of End-of-Life Vehicles (ELVs) in Europe points toward some answers.

## Design for disassembling: a non-competitive issue?[7]

In the early 1990s, Germany called the world's attention for its tough approach toward the implementation of Extended Producer Responsibility (EPR)[8] type of regulation, as a solution for post-consumer waste problems. In August 1990, the German Federal Ministry of the Environment (BMU)[9] drafted a regulation suggesting that car manufacturers should take back their ELVs without any cost for the consumer.

Attempting to cope with the new constraint, automakers took action in two complementary directions. First, they tried to convince legislators that suppliers of materials, parts and components, car dismantling and shredding companies should share the ELV responsibility, and that the cost of recycling should be borne in the market. In October 1990, the German automobile industry association (VDA)[10] responded to the Environmental Ministry with a concept for the future processing of ELVs. Second, automobile manufacturers started on a learning process concerning the ELV problem via pilot plants for dismantling. Even though political lobbying could delay the implementation of take-back regulation, automakers did not want to risk being caught by new cost elements without any practical alternative responses. Besides, at that stage, carmakers were also unsure whether the ELV management could eventually become a new source of competitive advantage.

The pilot plants were either established by single manufacturers or in cooperation with dismantlers and shredding companies, with the main aim of learning about disassembling times and the economic limits of recovering car parts. The results were disappointingly clear. Design and assembling techniques of automobiles made disassembling and recovery difficult and costly. Cars are made by many types of plastic composites that are difficult to detach from the bodies, and their identification demands costly time. To defeat these costs and difficulties, it became evident that it was necessary to involve suppliers in the design phase, working closely with them in order to overcome the main hurdles. Dismantling techniques were now to be considered integral activities in the life cycle of cars. Manufacturers needed to identify problems associated with not just the production and consumption of cars but also their disposal. Shredding processes received particular attention, since the increase in the number of plastic parts that were not disassembled meant that they were ending up in landfills, thus increasing costs for the business.

Questions about the recyclability of most plastic parts remained, but automakers started directing attention toward those components that were easiest to disassemble and most cost-effective to recycle. The creation of recycling networks via bi-lateral agreements with car dismantlers and shredding companies was the next measure taken by automakers in the period 1991–95. Although these networks represented a step forward to reducing ELV waste, they were restricted to a few car parts. As parts and components get smaller and require more tooling, disassembling becomes increasingly time consuming and costly, car manufacturers limited recycling to few big plastic pieces.

On the political side, already in February 1991 the VDA established a dedicated working group on ELV recycling called PRAVDA,[11] aiming at political and technical cooperation among its participants. Dismantling pilot studies were coordinated and results discussed among the members. Material recyclability was studied in close cooperation with the plastics, rubber and glass industries, and new dismantling tools, information systems and advanced material sorting techniques were developed. PRAVDA–VDA elaborated a common concept for the recycling of ELVs, which was more as a political stake than a technical document, since all automakers endorsed the proposal despite their preferences for different solutions at the technical level. Five years later, in February 1996, the BMU accepted the industry proposal for an agreement. Some recycling targets were established but they were obviously less demanding than the first draft regulation proposed by BMU in 1990. The German automotive industry had succeeded in its political efforts.

By the end of 1996, not only the German but also the French and Italian auto industry (the three major car manufacturers in Europe) succeeded in influencing the national governments to accept voluntary agreements as an appropriate strategy to deal with the 12 million or so ELVs per year in Western Europe[12] at that time. Given industry commitment, government promised to refrain from direct regulation. Overall, by signing voluntary agreements, the industry and national governments accepted the resolution that the automotive waste problem should be based on the shared responsibility of firms involved in the value chain of automobiles. The party would be over soon, though. The relatively comfortable situation carmakers achieved at home was counterbalanced by the plans of the European Commission (EC) to regulate car recycling. In July 1997, shortly after the German agreement was reached, the Commission presented a proposal for regulation.[13] According to representatives of the European Parliament, the national voluntary agreements, besides differing from each other, were based on conditions that weakened the potential of measuring their performance. After long negotiations, in September 2000 the European Parliament officially adopted the legislation,[14] even though it pushed back the date from 2003 to 2006 when producer responsibility applied to car manufacturers. Among other requirements, the directive required 85 percent of material recovery by 2006, of which 80 percent should be recycled; and 95 percent by 2015, of which 85 percent should be recycled. By 2002, European Union (EU) member states had to bring into force the laws, regulations and administrative provisions necessary to comply with the Directive.

### Complex products: when does it pay to reuse, recover or recycle?

The case of ELVs in Europe is exemplary in showing some key issues involved in the potential benefits of design for disassembling and product stewardship activities. The ambiguity involved in solutions for ELVs limited the capacity of governments to legislate, and the willingness of carmakers to develop independent activity systems to recycle their cars. As a result, the ELV issue became irremediably political. For legislators, the trade-offs between environmental impacts during the use and end-of-life phases of vehicles made it difficult to regulate. For instance, forcing automakers to use more steel (so to increase recycling rates) would result in heavier vehicles and therefore reduce fuel efficiency during use. Second, there are several ways of designing and managing the system for collection, treatment, reuse and recycling motor vehicles. As a result, the selection of one particular solution would be both controversial and eventually inefficient.

For carmakers, it became clear that car recycling was not a consumer concern. The costs of dismantling and low rates of reusable parts also made manufacturers to move away from setting individual recycling plants. Overall, they realized that, rather than sources of competitive advantages, ELVs would become a burden. As the case implies, the uncertainty about potential solutions for post-consumption waste makes it difficult to identify responsibilities and, in general, legislate about complex products, such as cars, home appliances and a wide variety of electric and electronic products. In the case of ELVs, the uncertainties were used by carmakers to transform the solution into a non-competitive issue and, therefore, share the costs of collection and recovery (for a distinction between competitive and non-competitive strategies, see Chapter 1).

In general, take-back regulations are prone to be controversial because, quite often, it is difficult to identify the private benefits resulting from it in an *ex-ante* basis. As it was highlighted in Chapters 1 and 5, consumers tend to see recycling efforts as intrinsic responsibilities of business and, in general, are not willing to pay for them. Product stewardship involving collection, selection, dismantling, reuse and recycling has a tendency to become non-rival. The take-back legislation for the Waste Electrical and Electronic Equipment Directive, known as WEEE Directive (2002/96/EC) is another example. Similar to what happened with car recycling, corporate strategies addressing the WEEE Directive seem to converge toward collective schemes set by industry players to recycle and process end-of-life products.[15] Although the legislation aimed at instilling some degree of Individual Producer Responsibility (IPR), industry collaboration and lobbying tries to move away from individual to shared responsibility. As a result, so far most product stewardship practices have become a license to operate in the industry.

Shared responsibility of product stewardship activities eliminates the incentives for individual producers to develop independent systems, which would eventually differentiate them from the rest. Collective treatment in post-consumption also eliminates the difference

between expensive differentiated products, and products that originally competed on low costs. This means that, at least for the end-
of-life phase, any eco-attributes generated via eco-design (in this case,
design for disassembling) are dissipated in the collective recycling system. As the cases of ELVs and electric and electronic products indicated,
most companies see the post-consumption activities as a burden to be
avoided and do not believe they may eventually generate any competitive advantages. The result is a self-reinforcing system in which
eco-design is often used as a way of avoiding regulatory burdens or
eco-activism. Generating competitive advantage out of product stewardship requires companies to create an activity system that is clearly
unique and independent from industry rivals. By adopting design for
dismantling and specialized systems for collection (using sales points,
for instance), recovery reuse and recycling can become cost effective
and eventually contribute to lowering the costs of new products. In
other words, the companies need to transform product stewardship
into a rival issue.

## CHANGING PRODUCTS' NATURE

Some companies look for advantages not by redesigning their products but rather by changing their nature. In some respects, this is what
Michael Porter would call product substitutes. The main logic behind
the biotech industry, for instance, is the substitution of petroleum,
a fossilized raw material, with renewable ones. In such context, *bio*
means renewable, which is often also less toxic and, more recently,
cheaper. The case of ethanol production in Brazil is exemplary.

### Bio-fuels: greener commodities[16]

In recent years, energy generation and bio-fuels have became the dar-
ling of green venture capital.[17] The sales of bio-fuels are expected to
reach US$72 billion by 2011 and many investors are eager to enter
this promising market.[18] Not surprisingly, the attention was directed
to Brazil, the country with both the most successful ethanol program
in the world and the highest potential for expansion. The Brazilian government established the ethanol program back in 1973, called
Proalcool (Pro-alcohol in Portuguese). The main goal of Proalcool was
to develop the local ethanol industry so to substitute a large portion
of petrol with bio-fuels[19] and to provide an alternative income stream

for sugar producers in the face of declining world prices for their commodity product. In 1979, the program was expanded by a new set of economic instruments and regulatory measures to increase the market penetration of ethanol-only vehicles. This was a novel approach at that time, and still stands as the most successful case of a large-scale alternative fuel program in the world. By 1986, 96 percent of new vehicles sold in the country were fuelled by ethanol.[20] After 1987, however, the price of oil receded, motivating the government to reduce gradually its support for the Proalcool program.

Although an official program to foster the use of ethanol was never reinitiated, the demand rapidly resurged in Brazil after the introduction of flex-fuel engines by car manufacturers in 1999. Flex-fuel systems allow a mixture of different fuel types (petrol and ethanol, e.g.) to be used without having to equip the car with separate fuel tanks, ignition systems or other components. Curiously, flex-fuel technology was developed with the initial aim of increasing the efficiency of internal combustion engines, so cars could meet toxic emissions regulations in Europe, Japan and the United States. Hence, flex-fuel technology was a by-product that could be offered at modest marginal cost to suitable car markets, Brazil in particular. The result was a remarkable comeback of the ethanol; 90.6 percent of the cars sold in Brazil were flex-fuel by 2008.

The exact amount of savings for the Brazilian government with Proalcool is controversial. There is no question, however, that the net result is positive.[21] By substituting imported oil with locally produced ethanol, the program clearly helped the Brazilian government with its international balance of payments. Internally, sugarcane farmers and producers of ethanol harvested most of the benefits but automobile manufacturers, fuel distributors and the consumers also gained. Besides the direct economic gains, Proalcool also generated wider social benefits and positive environmental externalities. As the program helped farmers to optimize sunk costs and promote investments in new distilleries, it indirectly promoted the creation of more than one million jobs in the rural sector. Even if the quality of these jobs can be questioned, some studies suggest that Proalcool was able to generate 152 times more jobs than oil production.[22] In addition to these positive social effects, Proalcool helped to minimize carbondioxide ($CO_2$) emissions in motoring, the most pressing ecological issue in transportation.

By 2007, Brazil annual production of ethanol was 17 billion liters per year, of which 14 billion were for domestic consumption. Although considerable, this level of production is dwarfed by future prospects. By 2020, the annual global demand for ethanol is expected to be situated between 50 and 200 billion liters.[23] Large sugarcane producers, such as Jamaica, Nigeria and India, as well as counties such Australia and Sweden, which have some experience with ethanol, can eventually grab a share of this market. However, the natural endowment and socio-economic profile of Brazil makes the replication of the ethanol program unlikely – at least at the same scale.[24] With the lowest production costs in the world, Brazil's share of this demand can be substantial. On the other hand, in order to meet demand and stay competitive, around US$100 billion is necessary to rework the distribution infrastructure and the efficiency of the sugarcane mills. Such makeover is essential to develop the export business for Brazilian ethanol, particularly to Japan but also to the US and Europe. The critical aspect, though, is security of supply. The major concern within these export markets is that Brazilian sugarcane producers may once again switch away from ethanol production if sugar prices rise, as happened between 1987 and early 2000s.

The landscape of the sector is definitely changing. The prospect of ethanol to replace oil for transportation caused a *gold rush* to Brazil. A multitude of newcomers is joining the traditional sugarcane farmers and their distilleries, which are still relatively small family enterprises. An early amalgamation in the industry resulted in a few dominant producers, the largest of which is Cosan with about 9 percent market share. Petrobras, the largest Brazilian oil company joined in ethanol production rather belatedly, accompanied by a few foreign companies. BP, for instance, acquired 50 percent of Tropical Energia, a local ethanol producer. Other incumbents include international private equity investment firms, such as Brazilian Renewable Energy Company (Brenco),[25] a start up funded by the American billionaire Ron Burkle and venture capitalist Vinod Khosla; large players in the agribusiness (e.g. Archer Daniels Midland and Cargill); global sugar producers and traders, (e.g. Sudzucker AG and Bajaj Hindustan); and private and public banks (e.g. HSBC, Inter-American Development Bank and World Bank).

This transformation process of professionalization, increased scale and internationalization is heavily dependent upon external capital, which tends to be volatile, as the repercussions of the financial crisis of 2008 best indicate. Nonetheless, the long-term trends toward de-carbonization and *peak oil*[26] suggest that the potential for profiting from bio-fuel production remains considerable. The opportunity relates to the nature of bio-fuels; that is, commodities with a lower impact than fuels based on petroleum. Bio-fuels have the intrinsic attribute of being renewable sources of energy and, because of this, are susceptible to lower regulatory risks than carbon-intensive fuels. Such attribute, however, is not prone to be valued as eco-branding products and services are. The marketplace does not reward the eco-friendly attributes of bio-fuels by paying price premiums, as niches for eco-differentiated products do. As with any other commodity, such as oil, corn or sugar, the central element is pricing. To be economically sustainable, ethanol production needs to be viable without the government subsidies that marked the early phase of the Brazilian Proalcool program. Indeed, it was necessary the costs of ethanol production[27] to fall from US$100 per barrel in 1980 to around US$25 in the early 2000s to became an attractive export commodity. Although the ecological attributes provide bio-fuels with a pole position in the race for cleaner fuels, the combination with low costs is what that makes winning a real possibility.

## Industrial biotechnology: back to nature, at low costs

The production of ethanol is not the only *industrial biotechnology* (or biotech, for short) example of Brazil. Dow, a giant chemical manufacturer is among a group of foreign firms investing in plants for the production of bio-ethylene. Traditionally, oil-based ethylene is the most widely produced organic compound, worth US$114 billion in global sales, which makes the investment in bio-ethylene a product for mass market, rather than niches for biotech products. The biotech version is very competitive simply because it is better, cheaper and more environmentally sound than the oil-based ethylene. Even with oil prices as low as US$50–60 per barrel, industrial biotech products such as bio-ethylene are still very competitive. As long as biotech products are technically superior and not just greener, they can enter the traditional petrochemicals markets. Companies such as Shell, BP, Dow and DuPont can blend biotech products into oil-based ones, which allows these firms to incrementally move into the new generation of bio-chemicals.

As the examples suggest, the embedded environmental attributes of biotech products are part of a major development currently underway in industry. The combination of corporate environmentalism, the approach of peak oil, advances in bioengineering, and the use of agricultural feedstocks to produce bio-polymers is driving the steady growth of biotech. Although biotech has not attracted as much attention as Genetically Modified Organisms (GMOs) or bio-fuels, it has been growing quietly for years. More than 20,000 patents are granted every year for biotech. In 2008, bio-products accounted for €300 million of sales of BASF, the German chemical giant, the same amount the Danish Novozymes obtained with the sales of enzymes for improving laundry detergents. Although remarkable, the examples are just the tip of the iceberg. The global market potential for industrial biotech has been estimated to reach US$100 billion by 2011.

The applied chemistry leading to industrial biotechnology is, in fact, not new. The transformation of agricultural feedstock into industrial and consumer products was initiated back in the 1930s by George Carver, who pioneered it by converting peanuts, sweet potatoes and other crops into glue, soaps, paints, dyes and other industrial products.[28] In that period, cellulose was already used to make paintbrushes and rolls of films. Henry Ford even developed car panels from agricultural materials, and built all-soy car panels back then. The World War II halted developments of industrial biotech, marginalizing the role of agriculture-based plastics, paints and textile fibers, among other industrial products. After the war, low oil prices and breakthroughs in petrochemical technologies ensured the dominance of petroleum-based plastics and chemicals.

In their early days, bio-plastics performed poorly, contributing to a bad image attributed to them. This aspect partially explains the reluctance clients still have to adapt equipment for bio-polymers, even when they are better or cheaper than the oil-based ones. A few companies are changing this. NatureWorks, a pioneering biopolymers firm, is a good example; it started as a research project within Cargill, a giant American agribusiness, then merged with Teijin Japan. NatureWork's bio-polymers are used in packaging material in products like diapers and juice bottles but could eventually be used in consumer products such as computers and mobile phones.[29] The bio-polymers, which generate 80–90 percent less carbon emission than traditional petroleum-based plastic packaging,[30] are marketed under the name of *Ingeo*. Such eco-attributes justify the great market potential of bio-plastics. Some expect that in the US alone, the demand for biodegradable plastic will reach US$845 million by 2012.[31]

Genencor[32] is another example of a fast growing industrial-biotech firm. The company is a division of Denmark's Danisco, one of the largest industrial enzyme manufacturers in the world. The enzymes are used in dishwashing and laundry detergents and in personal care products. Danisco recently expanded its business via a 50–50 joint venture with DuPont to form Dupont Danisco Cellulosic Ethanol (DDCE) to produce bio-fuels from non-food biomass feedstock such as switchgrass and corn stover and corn cobs.[33] On the other hand, DuPont expects that its new bio-fiber, sold under the name of *Sorona*, will be a multi-billion dollar product. The company expects the sales of biotech products to increase by 18 percent a year, to reach US$1 billion by 2012.

Although some early movers in markets for biotech products may be able to charge price premiums, industrial markets tend to be very cost sensitive. In general, interest in bio-polymers increases in tandem with oil prices. As a result, even though the demand for more eco-friendly products will play a major role in the purchase decisions, competition tends to converge on cost strategies. For companies operating in such markets, the deployment of E-cost Leadership strategies can represent a guarantee against regulatory measures and associated costs of carbon-intensive products. More importantly, firms that are able to manufacture products with embedded ecological attributes put themselves in a very competitive position, as the above examples illustrate. Some may go even further. By embracing the problems their products are supposed to solve for their clients, firms are eventually able to develop more eco-friendly products at lower costs.

## REDEFINING PRODUCTS' CONCEPT AND USE

Pastoral agriculture is the main farming activity in New Zealand. The temperate maritime climatic zone allows grass and clover growth over eight to twelve months a year.[34] The animals involved in pastoral agriculture are sheep (39 million), beef cattle (4.5 million) and dairy cattle (5.2 million), although more recent developments have included deer (1.6 million).[35] They are rarely housed during the winter but supplemental feeding of hay and silage in paddocks is common during the cooler months or at high altitudes. Animals are grazed in paddocks, often with electric fencing, which allows for rotational grazing and controlled pasture utilization.

When compared to most intensive-crop growing, pastoral agriculture has a lower environmental impact. But modern pastoral practices have intensified land use and lifted stocking rates, with the consequent increase in contamination of aquifers by nitrate – technically called nitrate leaching. Inorganic nitrogen fertilizers are often blamed for the problem but this is mostly unfounded. The application of urea, the most common nitrogen fertilizer used in New Zealand, contributes little to the total quantities of nitrogen leaching to ground water. Instead, it is the urine patch from cattle that causes the problem. The soil-plant system is unable to retain the high rates of nitrogen in the patch, resulting in the leaching of nitrate into groundwater.

Ravensdown, the largest fertilizer cooperative in New Zealand with 26,000 members, has been addressing the problems associated with intensive pastoral agriculture, dairying in particular. The company has been among the environmental leaders in the country working, for instance, to develop and update the Code of Practice for Fertilizer Use, introduced in 1998, and introduced nutrient management plans for farmers. In particular, the company has been very active in the Research and Development (R&D) of a solution for the problem of nitrate leaching. Through a partnership with Lincoln University at Canterbury (in the Southern island), various nitrification inhibitors and formulations at different rates and timing were tested in a three-year research program. The outcomes resulted in a commercially viable product, called *eco-n*, which increases pasture productivity by improving soil nutrient cycling in grazed dairy pasture.[36] *Eco-n* is a nitrification inhibitor sprayed onto dairy pastures in autumn and early spring reduces harmful nitrogen leaching and gaseous losses. Since the nitrogen remains in the soil, instead of leaching, it increases pasture yields.

By keeping a portion of the nutrients required within the soil system, *eco-n* does a similar job to nitrogen fertilizers but even if it contributes to higher plant yields, it cannot be considered a fertilizer *per se*. This is because rather than supplying additional nutrients, as fertilizers do, *eco-n* reduces the loss of nutrients already in the soil, deposited by grazing animals. Costs are reduced because farmers need to apply less nitrogen fertilizer. Hence, while traditional fertilizers work as *add on* to the supply of nitrogen, *eco-n* can be considered a *keep in* product for making the nutrient cycle more efficient. Overall, *eco-n* increases pasture yields while reducing both farmers' costs and the environmental footprint of pastoral agriculture.

Ravensdown patented *eco-n*. Even though the product has already obtained a first-mover advantage, securing patents in New Zealand it is crucial to prolong this situation and avoid direct competition in local markets. Interestingly, the company applied for a method patent, rather than for the actual active ingredient of the product.

The formulation of *eco-n* has been available for a number of years, but it had not been used in the context of spray application to pastures to deal with urine patches as *eco-n* does. The patent was first lodged with the Intellectual Property Office of New Zealand (IPONZ) in August 2002, approved by the examiner and announced for opposition in January 2004. But securing the patent has proven to be a drawn out process. Ballance Agri-nutrients, the main competitor of Ravensdown, challenged the patent validity, delaying its granting.[37]

Apart from the challenges in securing the patent, the combined economic and environmental benefits of *eco-n* resulted in a market success of the product. *Eco-n* started with just NZ$1 million (€457,000) or 10,000 Hectares (ha) in sales in 2004, but by the end 2008, the goal of NZ$10 million (€4.57 million or 70,000 ha) in sales was surpassed. Additionally, by 2013 farmers may qualify to apply for carbon credits, as a result of *eco-n* lowering emissions of nitrous oxide ($N_2O$) a greenhouse gas 310 times more potent than $CO_2$ from their farms.[38] For *eco-n*, it seems the sky is the limit.

## Multiple dividends of E-cost: product, place and planet

The case of *eco-n* is exemplary in showing the hidden advantages of an E-cost Leadership strategy: that is, a product presenting both the lowest costs and environmental impact. The embedded advantages relate to three aspects. The first one is the product itself. *Eco-n* was developed in a fundamentally different way. Instead of focusing on the nutrients to be added to the soil, as traditional fertilizers do, the product concept was based on readjusting the soil-plant system cycle under intensive pastoral agriculture. *Eco-n* components were defined not by comparing it with fertilizers but rather researching the fundamental problems fertilizers are supposed to solve. In doing so, Ravensdown found out that by developing a soil-management system based on a spray application of a new product, it would yield better results than applying traditional nitrogen fertilizers, besides partially offsetting them. In simple terms, by restoring nitrogen cycles, *eco-n* helps farmers to save money apart from protecting the environment.

Indeed, low cost has been central for the success of *eco-n*. Farmers may recognize the environmental benefits of applying the product but pastoral agriculture for dairy farming is extremely cost sensitive. Although *eco-n* created a new market of nitrification inhibitors for pasture application, it faces competition from products that either reduce the overall environmental impact of pastoral agriculture, or products that increase the supply of nutrients, as fertilizers do. Ballance Agri-nutrients, for instance, the second largest fertilizer industry player in New Zealand, developed *n-care*, a product that follows the traditional approach of using a nitrification inhibitor, that is, it is an add on product in fertilizers.[39] *N-care* aims more specifically at the effect the nitrogen fertilizer has in the loss of nitrates in the system, rather than on the urine patches, as *eco-n* does. *Eco-n* also faces indirect competition from other sources of nutrients or animal feed, and alternative farming systems that reduce nutrient losses. Overall, farmers will always measure the cost effectiveness of *eco-n* against alternative products and farming methods. They are simply not willing to pay price premiums for the ecological attributes of the product.

Nonetheless, eco-attributes do play a role. The optimization of nitrogen cycling results in better groundwater quality in the areas adjacent to the farm, hence reduced local impacts (place). Today, there is no direct reward to farmers for reducing the amount of nitrate ($NO_{3-}$) in groundwater. But this may change soon. Health standards for drinking water set maximum nitrate levels. In New Zealand, to safeguard human health, the Ministry of Health has introduced drinking water guidelines limiting nitrate concentration.[40] Local and central government are sending regulatory signals and some regional plans may require farmers to provide nutrient budgets, using approved modeling for nitrate leaching. Nutrient budgets will report annual inputs and outputs within a farm system and can be completed by trained staff, such as Ravensdown's field experts. Leaching is an inevitable consequence of land use intensification and current measures provide mitigation only, rather than a total solution to the problem. If these budgets show unacceptable rates of nitrate loss, then a plan to address this will be required. Here the use of *eco-n* becomes an additional asset. *Eco-n* is likely to be accepted as one of the methods of addressing these issues. The NZ Ministry for the Environment is preparing additional national standards that Regional Councils will have to adhere to. These may speed up the rate of change to a regulatory system that will open the way to increasing use of *eco-n* as a mitigation tool.

The third eco-attribute of *eco-n* relates to global impacts of agriculture (planet). Nitrous oxide ($N_2O$) is a powerful Greenhouse Gasses (GHG) and one of the culprits for the NZ agricultural sector to be accounted for 50 percent of New Zealand's carbon emissions. Nitrous oxide makes up one third of agricultural emissions with two thirds derived from methane from ruminant animals. New Zealand ratified the Kyoto Protocol in 2002 and, as part of its Climate Change Policy Package, proposed to introduce a carbon tax, which was abandoned after the national elections of 2005. Abandoning the carbon tax means other options for greenhouse gas mitigation need to be introduced by the government, including market-oriented mechanisms and technologies such as the ones embedded in *eco-n*. Even if uncertain, the prognosis for *eco-n* is very positive, since it offers an important agricultural mitigation technology. If the product were to be used by all the country's dairy farmers (1.4 million ha), it would reduce the annual cost of nitrous oxide from agriculture by approximately NZ$40 million (€18.6 million), based on emission levels and carbon pricing in late 2008.[41] The NZ government urgently needs to find ways to pay for carbon deficit, and the environmental gains brought by the *eco-n* technology can emerge as a viable option.

In conclusion, E-cost Leadership presumes that companies compete on the basis of cost, with the environmental attributes of the product representing either the license to operate in cost-sensitive markets or, as the case of *eco-n* imply, additional sales arguments. New Zealand farmers may buy the *eco-n* (product) simply because it reduces the overall production cost of dairy farming. However, by applying *eco-n*, the cycles of nitrogen will be optimized, resulting in lower levels of groundwater contamination by nitrate, allowing farmers to be better prepared to comply with more restrictive water-quality regulations (place). Finally, the reduction of nitrous oxide ($N_2O$) emissions may entitle farmers to carbon credits in the near future, representing an additional value of applying *eco-n* in their pastures (planet). Although Ravensdown will not be able to obtain price premiums for *eco-n*, the eco-attributes of the product represent additional sales arguments, allowing the product to be an E-cost leader. This is good news for the owners of *eco-n*, but what about products that have little room to change their formulation? Can they compete on the basis of E-costs? What is necessary for them to reduce their overall environmental impact? The case of Chemical Management Services (CMS) is exemplary in showing how this is possible.

## Aligning buyer-supplier interests via servicing

Complex supply chains for chemical products often result in suppliers not knowing how their products are used and the environmental damage caused by them. Quite often, industrial clients also have little knowledge about ingredients, and how to use and dispose the chemicals. Such lack of knowledge creates uncertainty about their final destination and associated costs, as well as health and environmental risks. The provision of CMS emerged in the 1990s as a way of addressing such problems. In a CMS, the supplier has the responsibility to manage the chemicals and reduce their costs for the user.[42] In other words, CMS alter the logic of traditional buyer–supplier relationships, where the buyer tries to sell as much of a product as possible, and the buyer tries to buy as little as possible. With CMS, both parties have the same incentive to reduce costs and eliminate wastes.

The Chemical Strategies Partnership (CSP),[43] a non-profit organization based in San Francisco, California, has been promoting the diffusion of CMS. Among several cases presented by the Partnership, a long-lasting example is the *pay as painted* contract between PPG, a supplier of chemical services, and Chrysler, the automaker. Since 1989, PPG provides services for body surface preparation, treatment, and coating chemicals, and owns the chemical until they are used. Considering PPG is not paid until the car is produced, it has a vested interest in reducing the amount of paint used in each car. According to the CSP, Chrysler saved US$1 million after the first year, besides reducing Volatile Organic Compounds (VOC) emissions.[44] Again, since PPG is paid by painted car instead of gallons, the company is interested in the overall efficiency of the system. In other words, the reduction of consumption of chemicals is beneficial to both the supplier (PPG) and the client (Chrysler). Overall, this is expected to reduce both the costs of the product/service *and* its overall environmental impact; an unambiguous E-cost strategy.

Northern-European companies[45] noticed several advantages in adhering to CMS. Among them are higher degrees of security in the management of chemicals. By reducing the use of chemicals, factories can reduce costs, emissions and the exposure to liabilities. Since the supplier does not try to sell as much as possible but only charges for the quantity used, another benefit of CMS is the virtual elimination of waste in the factory. CMS also result in a stronger relationship between the client and the manufacturer, which often leads to process development, optimization and eco-efficiencies. However, European companies also have some concerns over CMS, such as uncertainty about the responsibility in the case of an accident. Exposure to trade secrets is another issue, since close relationships between the manufacturer and the clients may facilitate access to manufacturing technologies and commercially sensitive information.

## Product-service systems: redefining the income basis

As the CMS case suggests, by shifting from selling products to selling the functions provided by them, some firms operating in industrial markets (B2B) can reduce both economic costs and environmental impacts. Differently from the case of biotech products, the ingenuity here lies in the possibility of reducing environmental impacts without fundamental changes in the nature of the product. After all, when product improvements reach a limit, further reductions of costs and environmental impacts can be achieved only by changing the nature of the buyer–supplier relationship. For instance, if all chemicals used for cleaning a factory were eventually substituted by biotech ones (i.e., made from renewable sources), we could expect a reduction of the overall environmental impact and, eventually, costs. Further environmental and costs savings, however, can only be achieved by reducing the total amount of material used in the factory. For that, changes in the nature of the buyer–supplier relationship are required, as the CMS best illustrates. By redefining what is sold (painted cars, rather than paint *per se*), suppliers and buyers have the same interest of trimming down the quantity of products used in a certain process as much as possible. The result is a reduction of both costs and environmental impacts.

Such types of commercial affairs fall under the category of Product-Service Systems (PSS).[46] By selling the functions products are supposed to deliver, revenues can be created at both lower environmental impacts and costs. Not surprisingly, such win–win scenario has appealed to environmentalists, Non-governmental Organizations (NGOs) and international organizations such as United Nations Environment Programme (UNEP),[47] as well as to academics who saw on PSS the possibility of decoupling the overall economic growth from the materiality of the economy.[48] The enthusiasm is valid: deployed at large scale, PSS could lead to lower environmental impacts and, eventually, to sustainable development. This rationale led UNEP and other sponsors to promote research and empirical development of PSS[49] since the early 2000s. Among the cases studied in the past years are communal washing centers, remanufacturing business models, such as the one of Xerox photocopiers, the carpet leasing program of the Interface, an American corporation, and car-sharing systems (explored in detail in Chapter 7).

The expectations that PSS should be widely diffused, as well as normative arguments supporting its deployment reflect a utilitarian (or functionalist) view of consumption, which is anchored in neo-institutional economic theories.[50] The functionalist perspective assumes that, when deciding about different alternatives, rational cognition is the main driver of consumer behavior. As long as economic advantages emerge, consumers would be willing to opt for services rather than product ownership. Such logic should lead to the widespread adoption of PSS but the reality shows otherwise. Empirical evidence of PSS is disappointingly scant. We are left, then, with some obvious questions: If the economic benefits of PSS are so great for all parties involved as well as for environmental benefitis, why are there so few empirical examples?[51] Why only a few companies have moved their revenue basis from products to services? Overall, when does PSS pay?

Broadly, the main barrier to adopting PSS is cultural. Although PSS works well in some specific areas of industrial markets, as the case of CMS suggested, cultural and legal sensitivities, as well as management discretion limit broader applications. Sometimes, the issue might simply be pride. People involved in manufacturing are often overconfident of their competences in managing resources. Allowing external parties to work inside their facilities can be more than a cultural shock; for some, it might signal incompetence.

There are, obviously, more objective reasons limiting the participation of third parties in the management of factories, retailing or other business-to-business (B2B) activities. Quite often, it is not easy identifying the economic gains brought by external services such as PSS. Accounting systems are often complex, and isolating costs by activity can be daunting.[52] Even when it is possible, legal systems and taxation rules may limit leasing services (a way of deploying PSS), as Interface encountered with its American clients.[53] The account aspect leads us to management discretion, another limitation for the move toward servicing. Similarly to what happens to the treatment of wastes, mentioned in Chapter 3, managers' definition of the core competence of the business significantly reduces their strategic choices. For instance, if carmakers see manufacturing car bodies and engines as their core competences, it is very unlikely that they will consider servicing car-sharing systems as part of their business. Besides, moving from products to services requires a broad shift in competences, which most manufacturers do not have or simply prefer not to bother with.

Finally, not only companies are entrenched in their views of what constitutes their business. Markets are realms of significations in which consumers have ingrained images about products. For many of us, products are *symbols for sale*[54] that go far beyond functions, as emphasized in Chapter 5. Although functionality obviously plays an important role in product performance, reducing consumption to its instrumental facet is to ignore the emotional dimension of human intelligence. After all, not by chance marketing and advertising emerged as both prominent industries on their own right and sciences strongly anchored in psychology. For PSS to prevail as many wish, there is the need to reinforce the symbolic meaning of *usership* at the expenses of ownership.[55] For that, it is often necessary to refocus on branding efforts, which may result in higher costs, possibly requiring service providers to adopt Eco-branding or Sustainable value Innovation (SVI) strategies (presented in Chapter 7), rather than the E-cost strategy. By adopting a dedicated service brand, service providers may be able to retain, at least partially, the symbolic meaning attached to ownership. Overall, acknowledging the symbolic component of consumption is fundamental to explain the relative scarcity of empirical examples of PSS and its potential to succeed in the near future. Although there is surely scope for the expansion of PSS, similar to the areas prone for E-cost leadership strategies, there is the need to identify when such systems have the best chance to succeed. The following section provides insights into this direction.

## WHEN *E-COST LEADERSHIP* PAYS

Broadly, companies that need to compete with product portfolios on the basis of low cost, as well as reduced environmental impacts need to focus on E-cost leadership strategies. To do so, however, they need to observe key elements relating to the context in which they operate, which includes the nature and attributes of the products and their target markets. They also need to create the necessary organizational competences to become the E-cost leaders.

### Context

In principle, there is scope for E-cost leadership in almost every sector. After all, in most markets, there is always a space to focus on low costs. But firms operating in industrial markets (B2B) are more prone to profit from E-cost Leadership strategies than those in consumer markets (B2C). These firms tend to face ever-tightening environmental regulation and have little scope for differentiation, which forces them to compete mostly on the basis of low price. For instance, Boing and Airbus, the leading makers of jumbo jets, may be willing to pay higher prices for a jet turbine that causes less noise and consumes less fuel. After all, in the past decades there has been ever-tightening regulatory demands for airplanes to be quieter and cleaner (less emissions), while the rise in fuel prices has pressured air carriers to cut costs. Hence, a cleaner, more efficient and quieter turbine (GE or Rolls Royce, for instance) is certainly welcome by Boing or Airbus. But the willingness to pay higher prices to the supplier of the turbine is normally coupled with cost cuttings during the life cycle of the product. In the end, as it was explored in more detail in Chapter 5, in industrial markets pricing tends to be associated with potential cost savings down the line. The result is a more rational marketplace, when compared with consumer markets.

As the case of the Ecolean suggested at the beginning of this chapter, companies supplying relatively simple products, such as packaging and raw materials for other business can obtain competitive advantages when able to reduce the environmental load of products, provided they are cost leaders first. Recognizably, a few players in industrial markets may be able to obtain price premiums via eco-differentiation, but for the majority of suppliers, pricing is a chief determinant of sales. This is also the case for the vast majority of industrial biotech products, which includes bio-fuels and a wide range of bio-polymers. Even though these products have embedded eco-attributes, clients and consumers are not able or willing to pay higher prices for them. Exceptions do exist, in particular in some applications of bio-polymers, but in most cases biotech products are commodities that have their prices directly or indirectly determined elsewhere (at the Chicago Board of Trade, for instance). The nature of these products – of being carbon neutral, biodegradable, etc – provides them with the license to operate in new markets, such as the one of ethanol for fuel or bio-polymers for packaging, but competing within these markets is primarily done on the basis of low costs. Hence, for the specific case of E-cost leadership strategies, the return on eco-investments depends on conceiving or designing products in a way that they are able compete in markets under heavy regulatory demands (packaging, for instance), as well as enter new biotech markets, such as bio-fuels and bio-polymers.

Businesses tend to transform product stewardship activities into a non-rival issue, as the cases of take-back legislation for ELVs and Waste of Electric and Electronic Equipment (WEEE) in Europe suggested.

Shared responsibility of activities associated with recycling eliminates the incentives for individual producers to develop independent systems, which would eventually differentiate them from the rest. In complex products, because of the intricacies involved in the collection and processing of post-consumption waste, eco-design has so far helped companies to reach the targets imposed by regulation; very seldom it helped them to generate competitive advantages. For instance, by applying eco-design principles and tools, carmakers and manufacturers of mobile phones and electric home appliances are better prepared to reach the recycling targets imposed by the take-back EU Directives.

Hence, by adopting basic eco-design requirements, manufacturers avoid the risks of not reaching the imposed recycling targets. However, if they adopt shared responsibility of post-consumption waste, no competitive advantages will emerge from product stewardship practices. Eco-design practices can generate competitive advantages only when companies are willing or able to move away from shared responsibility and work toward post-consumption practices that are unique to the firm. For instance, by adopting a different business model, a supplier of computers may create specialized centers for collection and recycling that deals exclusively with its own products, hence facilitating disassembling, reuse of parts and recycling. Once the company creates a different set of activities for the treatment of its products, it will eventually *be better by being different*.[56]

Companies can also reduce costs of traditional products by adopting innovative business models, such as the ones required for Product-Service Systems (PSS). By moving from selling products to selling the function provided by them, suppliers and buyers can align their interests and trim down the amount of products used in a certain process, as the CMS case illustrated. Although great in theory, in practice PSS has encountered hurdles. Besides the broad cultural context in which consumption is attached to the private ownership, PSS is limited because most applications do not compensate clients or consumers for the loss of control resulting from the move from product ownership to services. Such loss can be more or less objective or symbolic. In industrial markets, most companies do not want to give away control of certain activities performed in house due to a series of discretionary management issues. In consumer markets, the loss can be more symbolic than functional. Most of us are not willing to trade the certainty of product ownership for the evanescent nature of servicing, which can vary in quality and convenience, often at reduced status.

In order to circumvent the hurdles for the adoption of PSS, companies may have to deploy branding strategies (discussed in Chapter 5), which may result in higher costs and, therefore, limit the application of PSS as a low-cost strategy. In order to succeed in consumer markets, PSS needs to regain the intrapersonal symbolic qualities that often disappear with the loss of ownership, as well strengthen the interpersonal qualities that may have been diluted during the transition from products to services. Overall, specific condition facilitated PSS to be deployed as CMS in the US and Europe, and as communal laundry services in Sweden or car-sharing in Switzerland. Such examples, however, seem to be the low-hanging fruits of PSS. The diffusion of this type of business model requires more than a utilitarian view of commerce. Besides the symbolic dimension of PSS, it is necessary systemic changes in the context of both production and consumption, as Chapter 7 explores in detail.

## Competences

Lowering economic and environmental costs simultaneously is a challenge. First, it requires managers to adopt life cycle thinking about their products, which, in the majority of the cases, will normally result in higher levels of managerial complexity. In order to accommodate requirements for lower impacts, products may have to be designed in a fundamentally different way, including, for instance, the substitution of raw materials, elimination of toxic components, as well as easing disassembling, reuse and recycling. Quite often, such requirements result in higher costs, demanding companies to innovate products as well as the way they are used or consumed, so costs can be mitigated or reduced. In order to become leaders in both reduced costs and environmental impacts, companies have to develop a wide set of capabilities. The following questions can indicate the most critical ones:

- What products (or line of products) in our portfolio are required to present ever-increasing environmental performance while only competing on costs?
- What is necessary to substantially reduce the environmental impact of these products? Can we do it while maintaining the cost leadership in that segment?
- Do we have competences in Life Cycle Assessment (LCA)? Do our designers have a clear understanding of the environmental impacts

of our products throughout their life-cycle? Do they have a good understanding of eco-design principles? Are they trained to employ such principles in the re-design of our products so to reduce both environmental impacts and costs?

- If any of our business clients requests us to present a product substantially easier to disassemble, are we prepared to deliver? Can we increase the rates of reuse and/or recycling of our products by adopting design for disassembling techniques?

  - How difficult is to disassemble our products? Do we have any idea about the time necessary for total disassembling?
  - How many parts can be reused or recycled?
  - Can we alter anything in the configuration, materials and type of fastener of our products so to ease disassembling at lower costs?

- Can we reduce the weigh, toxicity and overall environmental impact of our products by using alternative materials?
- By changing the design of our products, can we substantially reduce the number of components so to reduce both costs and environmental impacts?
- Can we make our products reusable? What would be necessary for that? Would such move increase or decrease our revenues? Are our clients willing to accept refurbished products?
- Can we redesign our products in tandem with the design of their packaging so to reduce weight, volume while increasing reuse and recycling rates at lower costs?
- If regulators demand us to collect and recycle our products, should we partner with our competitors, so to transform recycling in a non-rival issue, or should we strive to create our own recycling network, with the aim of reducing cost and generating competitive advantages?
- Can we substantially reduce the environmental impact of our products by changing the way we sell them? Can we transform our products into PSS? What would be necessary from us and from our clients? Are our personnel prepared to become service providers? Do we understand what is involved in supplying services? Do we have a trustworthy relationship with our clients so to propose to work inside their factories?
- Do we understand the symbolic dimension attached by customers using our products? Can we transfer part of this symbolism from ownership to *usership*? How can we strengthen the intra and

interpersonal symbolic qualities of our products when moving to servicing?

- By using our products, are clients able to generate carbon credits? If so, what can we do to help them to obtain such credits?
- How can we deter our competitors from imitating our products or business models? Can we protect them via patents?


## CONCLUSION

Companies capable of exploring E-cost Leadership strategies put themselves in a superior competitive position in existing markets. They satisfy the basic and often most important requirement of commerce: low costs. Independently on whether clients or consumers value the eco-attributes of products, firms are able to compete in markets with razor-thin margins. Obviously, this is a very good position to be in, since E-cost leaders are also prepared for increasingly demanding requirements for eco-friendly products, either from regulators, consumers or both. If this is simple in theory, however, is very hard to be done in practice. This is the main reason why a wide range of practical examples were used in this chapter to illustrate the ways by which firms can lower the environmental burden caused by their products while keeping costs down.

There are several ways of reducing the intrinsic-environmental impact of products. Among them are dematerialization, product substitution and a series of additional guidelines, methodologies, tools and techniques under the concept of eco-design. In certain cases, the eco-attributes of products can be used as sales arguments, in addition to low costs. As the examples used throughout the chapter suggested, eco-attributes have helped companies to cope with regulatory requirements while helping them to compete in low-margin segments. Eco-design helped packaging manufacturers, for instance, to reduce the environmental load of their products while also reducing costs. Some companies went even further and changed the nature of the raw materials or the entire products, so to maintain their presence in traditional markets (bio-polymers blended into petrochemicals, for instance), or access fast growing eco-oriented markets such as the one of bio-fuels. Hence, competitive advantage in the context of E-cost strategies results from the company being able to be the leader in existing cost-sensitive markets or access new ones that require the products to present eco-attributes. In other words, the eco-attributes of the product may become the license to operate in some cost-sensitive markets.

Some of the cases discussed in this chapter also hint the possibility of companies to move from the realms of existing industries to new market spaces. The case of *eco-n*, discussed in the chapter, is exemplary. The application of *eco-n* in grazing pastures optimizes the nitrogen cycles in the soil-plant system, reducing the need for nitrogen fertilizers and lessening the overall impact of pastoral agriculture. At the first sight, *eco-n* seems to have created the new market space – of nitrogen inhibitors. A closer look, however, suggests that creating of a new product segment (of nitrogen inhibitors) within the broader market for products and services within pastoral agriculture. Although *eco-n* does not have any direct competitor, it faces indirect competition from alternative products and farming techniques that, combined, do a similar job. In this case, the product eco-attributes help farmers to reduce the total costs associated with pastoral agriculture, which is a great asset for the product to compete in such cost-sensitive market. Hence, *eco-n* is a typical case of E-cost strategy, but the case indeed hints that, by looking into the ultimate problem the product is supposed to solve (optimized nitrogen cycles, in the case of *eco-n*), companies may eventually be able to enter unexplored territories.

Some examples of Product-Service Systems (PSS) such as the Chemical Management Services (CMS), besides making a clear case for E-cost leadership strategies, also hint the possibility of moving companies away from the fierce competition within existing industries. By changing the terms of the contract between buyers and suppliers, PSS can result in environmental and economic gains. However, in order to move consumption from products to services it is necessary to deploy new business models, and associated activities, as well as broader systemic changes. This may result in a new value proposition. As the CMS indicated, PSS may even result in value innovation – in which additional value is created at lower costs. While the concept of PSS is very promising, the chapter explored the reasons for relatively scarce empirical applications. Among them, besides managerial discretion and symbolic elements associated with product ownership, are the required systemic changes for the successful deployment of services. Quite often, systemic changes extend to infrastructure and collaborations among a new set of players.

Broadly, PSS define the borders between the four Competitive Environmental Strategies (CES) and Sustainable Value Innovation (SVI), presented in Chapter 7. In particular, E-cost and SVI strategies share a few elements, but there are decisive differences between them, pricing being the main divisor. While E-cost strategies are constrained by price rivalry (after all, the product must be the price leader), SVI creates new market spaces in which price comparisons are difficult or less relevant. Hence, pricing in SVI is not a constraint as it is in E-cost strategies. Additionally, the value proposition of SVI tends to go further than the straightforward eco-attributes embedded in E-cost strategies. As Chapter 7 explores in detail, radically new value propositions capable of creating not only private profits but also satisfy social and environmental demands may result in value innovation, which is both ecologically and economically sustainable.

# PART III
## BEYOND COMPETITION

# 7

# SUSTAINABLE VALUE INNOVATION

The previous chapters presented the four Competitive Environmental Strategies (CES) available to corporations to compete in existing industries (Part II). This chapter presents the fifth sustainability strategy, which builds on the concept of value innovation. According to the logic of Blue Ocean Strategy (BOS), by creating additional value to customers at lower costs, companies can bypass the competition of an existing industry because such value innovation can generate new markets spaces or, to use the BOS metaphor, "a blue ocean where the company can swim alone". BOS also presents a subtle caveat: As long as value innovation is created to existing or previously neglected non-customers, environmental impacts resulting from the new offer do not restrict its deployment.[1] On the other hand, as the chapters of Part I suggested, *Sustainability Strategies* entail embedding the value proposition of the company (private profits) into the broader environmental and societal context (public benefits). This means that the creation of Sustainable Value Innovation (SVI) requires companies to lower costs and increase consumer value while generating public benefits in the form of reduced environmental impacts and value for society.

As the examples in this chapter demonstrate, SVI can be created by redesigning the activity systems involved in both production and consumption of goods and services. In the majority of the cases, SVI strategies are possible only by redefining how products/services are produced and consumed. This is why SVI can be considered a systems strategy, for it requires changes not only on the nature and technology of products but also in the logic by which systems of production and consumption are organized. By questioning the appropriateness of the business model in creating value for shareholders and consumers, as well as for society, SVI strategies redefine the boundaries of the value system of an existing industry.

## THE CALL FOR SVI STRATEGIES: THE TROUBLESOME AUTO INDUSTRY

The economic, environmental and social impacts of the automobile industry are extraordinary. Alone, the industry is responsible for 50 percent of the world's oil consumption, 50 percent of the output of rubber, 25 percent of glass, 23 percent of zinc and 15 percent of steel. Altogether, it represents around 10 percent of the GDP in rich countries. The American car industry, with around 15 million cars per year, burns 8 million barrels of oil per day. The industry is responsible for 25 percent of United States (US) greenhouse gases and creates 3.2 billion kg or unrecycled scrap and waste every year.[2] Globally, the fleet is around 700 million vehicles, with expected 1.3 billion vehicles by 2020. More than 53 million cars[3] and 21.5 million trucks were produced in 2007, with an estimated value of the automotive value chain at above €1 trillion.

Throughout the 20th century, the industry played a decisive role in the economic growth of many nations. Car manufacturing has stirred the industrialization process in countries such the US, Germany, Japan, Italy, France, Korea and even Brazil, which does not even have an indigenous car industry. The industry mastered the technologies assisting the manufacturing of the modern car. Internal Combustion Engines (ICE) have achieved an amazing degree of sophistication and efficiency[4] and the technology associated with the production of car bodies made of steel has been the main driving force in the development of robots in factories. Not only have an impressive array of technological developments been achieved during the history of the automobile but the sector has also served as a benchmark for new management strategies and techniques, such as Total Quality Management (TQM) and Lean Thinking (LT) (mentioned in Chapter 3).[5] Terms such as *Fordism* and *Toyotism* became associated with specific management techniques, widely adopted by organizations in other industrial sectors.[6]

Not surprisingly, car-making also served as a test-bed for Blue Ocean Strategies (BOS). In their best-seller book, W. Chan Kim and Renée Mauborgne presented an "historical pattern of blue ocean creation for the automotive industry" – for the early days of the industry until late 1990s.[7] In this respect, this chapter builds upon and expands the logic of value innovation presented in *Blue Ocean*. In order to identify sustainability strategies for the industry, as well as for any new player in the area of Terrestrial Individual Motorized Mobility

(TIMM)[8] the following sections present the most recent strategies adopted by carmakers to face the demands for environmental and social responsibility.

There is a straightforward reason to use the motor industry as the backbone for the chapter. Considering the size and impact of the automobile sector in economic, social and environmental terms, "if this industry can fundamentally change, every industry can."[9] Hence, the chapter uses the car industry as an all-encompassing example, complemented by research cases to illustrate the various characteristics that make the SVI a strategy particularly suited to situations in which systemic efficiency improvements are not only economically feasible but also desirable from both environmental and social aspects. Since "an example speaks for one thousand words", the automotive industry is instrumental in demonstrating how companies under high levels of pressure can generate long-term value innovation by simultaneously addressing economic, environmental and social demands.

The following sections analyze the industry and present SVI strategies that can be explored not only by carmakers but also by new players in the *organizational field* of the automobile – key suppliers, consumers, regulatory agencies and other organizations that produce similar services or products.[10] Indeed, players outside the realms of the car industry, who are still relatively small, have initiated most cases presented in this chapter. When compared with the overwhelming dominance of the private-car ownership, such players can even be considered marginal businesses. Nonetheless, such comparative order of magnitude should not blind us to the real potential of SVI strategies. Although skeptics may see the scaling up of business for individualized mobility too radical or improbable, the evidence that most players in the auto business are not only ecologically but also economically unsustainable, justifies the search for alternative business solutions, which have been exploited by outsiders of the industry. For some car manufacturers, pursuing a long-term SVI strategy is less a question of bold management than the last chance before takeover of bankruptcy. For others, deploying a SVI strategy is a reasonable alternative simply because a radical mid-course correction does not require a scientific breakthrough. Dominating existing markets or, more importantly, creating new ones, depends mostly on managerial audacity, rather than on technological development. Indeed, this is the most compelling aspect of a SVI strategy for individual motoring: Because the modern automobile system is so inefficient, economic, environmental and social gains can be achieved at large scale with existing technologies.

Other industries have their idiosyncrasies and, obviously, require a specific analysis for the identification of SVI. Nonetheless, the underlying logic used in the study of the individual mobility business can definitely serve as a benchmark for other sectors. For any industry or business, SVI strategies require an evaluation of the efficiency of the system of production and consumption of a particular product, group of products and/or services. In order to identify systemic inefficiencies and leverage points for a radically innovative strategy, the study has to consider how product concepts and designs impose specific technologies of production, as well as particular patterns of utilization. As the following sections indicate, in order to create SVI, it is necessary to address systemic inefficiencies.

### The auto industry and economics: *red oceans*

In the past two decades, the foundations for the historic success of the car industry started to fade away. Remaining profitable or exiting from the business without substantial losses became increasingly difficult, with the majority of *volume*[11] car manufacturers frequently reporting significant financial losses.[12] As a result, the imperatives of cost reduction have been extended far beyond the frontiers of the focal car corporations. As signs of economic fatigue and the pressure to reduce its environmental impact escalate, the industry entered a profound process of restructuring. Improvements have indeed been made but, as a whole, the industry is in crisis, facing one of the most enduring challenges in its history. From the economic perspective, it is clear that recent strategies of large car manufacturers,[13] such as Ford and GM are not producing the expected results.[14]

Manufacturing automobiles has become an increasingly risky enterprise. The imperative of economies of scale have often caused volume car manufacturers to swing between profit and loss. For each model, a *break-even point*[15] between 100,000 and 150,000 units per annum is considered a minimum condition to be profitable in the industry. On a factory basis, a typical plant with a capacity between 250,000 and 300,000 units per annum must reach 65 percent utilization to be viable.[16] Slim margins also reflect the failure to capture the life cycle profit streams generated by car sales, ownership and use. From all activities involving car manufacturing, retailing, leasing, servicing, insurance, finance and car parts, among others, carmakers take a meager one percent of the total.[17] In terms of operating margins, car assemblers and component suppliers together have average returns on revenue at 3.5 percent. The combination of such low margins with high break-even points for most car models makes the business vulnerable to slight market fluctuations; the saturation of markets in highly industrialized countries transforms this vulnerability into a cycle of profit and loss. During the 1990s, for instance, only BMW, Toyota and Honda did not experience any year of losses. Over the period 2005–2006, GM and Ford, along with their primary suppliers Delphi and Visteon all reported financial losses, while in Europe, MG Rover finally collapsed into bankruptcy.

The expectations that emerging markets would work as a buffer zone for excess capacity encountered a tough reality. Problems common to most developing countries, such as a lack of infrastructure, currency fluctuations and consequent price instabilities and structural problems resulting from inequalities in income distribution, limited the market projections to become reality. For instance, during the second half of the 1990s the projection of sales of motor vehicles in Brazil for the year 1998 was 2.5 million units. Such a prognosis was one of the reasons for the massive influx of automakers' investments during this period, which transformed Brazil into the country with the highest number of local market vehicle producers. The predicted market sales only became a reality ten years later. By then (2008), however, Latin America presented an overcapacity similar to those of mature markets.

The financial crisis of 2008 only accentuated old problems faced by the sector. The automotive industry in Western Europe, Japan and North America have long presented quasi-horizontal curves as a result of both overcapacity and market saturation.[18] Higher levels of rivalry among vehicle manufacturers favored consumers to request a wider variety of models at lower prices. This market fragmentation puts intense pressure on a manufacturing system orientated toward standardized production and, alongside shortening model life cycles, undermines per-model economies of scale, increasing the significance of cost-reduction measures. In competing for an increasingly fragmented[19] market share, vehicle manufacturers are pressured to invest in product development, while low profitability requires them to cut costs. The result is daunting. The combination of the imperatives of economies of scale, market saturation, high consumer demand for a wider range of models, and low-profitability margins, has transformed the car industry into a *bloody ocean* of rivalry. In addition, as the next section delves into, the industry has been under relentless pressure to improve its environmental performance. Finally, on the social aspect, there are no indications that the sector is directly contributing to raising the standards of living of underprivileged social classes – indeed an emerging demand of Corporate Social Responsibility (CSR). From whatever angle, the pressure to change in the auto industry is overwhelming.[20]

## The auto industry and the environment: *cloudy skies*

The scope of environmental harm caused by cars is vast. The automobile constitutes an example of a product with an extensive environmental footprint in all phases of its life cycle,[21] as well as with related systems, such as road and supply infrastructures. According to the German Environment and Forecasting Institute, before an average car is put into use, it has already produced 26.6 tones of waste and 922 cubic meters of polluted air.[22] Although this is a remarkable figure, it represents less than 10 percent of the total environmental impact of an automobile during its life cycle. About 80 percent of its impact results from air emissions during car use; the remaining 10 percent is due to the pollution associated with its final disposal.[23] Although the modern car is remarkably efficient in terms of its adaptation to a specific technological option (all-steel monocoque bodies and ICE),[24] it is extremely inefficient in the extent to which it converts resources into individualized mobility.[25] Due to losses within the Internal Combustion Engines (ICE), the transmission components, as well as due to the heavy steel body, a typical modern sedan uses 70 percent of the fuel to transport its own mass, loosing further 12 percent in aerodynamic and rolling friction and in breaking. If just one person occupies the car, it means that out of 100 liters of fuel, 97.6 transport the car and just 2.4 liters, the driver.[26] Despite more than one hundred years of development, the modern automobile is remarkably inefficient.

Since the 1970s, carmakers have managed to reduce significantly the emissions of particulates and toxic gasses, and the average environmental performance of most fleets has significantly improved. However, as regulatory measures addressing climate change become politically charged, governments will push emission reductions to limits only achievable by alternative powertrains. The European Union (EU) intention to request an average fuel efficiency of carbon dioxide ($CO_2$) emissions of 130 g/km by 2012 is just an indication that the skies will be increasingly cloudier for carmakers. As in the recent past, the industry will further improve the efficiency of burning hydrocarbons in ICEs, but the laws of physics and chemistry establish a limit for reductions. In a decade or so, in many parts of the world complying with regulations will require carmakers to move away from ICEs. In times of low-profit margins, breaking away from ICEs will not be an easy task. In order to operate in a carbon-constrained world, satisfying standards of environmental performance will require automakers to invest ever-increasing amounts of money in increasingly expensive R&D activities. For many, the eco-investments in low- emission technologies, described next, will be done under very cloudy skies.

## THE DOMINANT STRATEGIC LOGIC: *GREENER CARS*

How can automakers satisfy environmental, market demands and remain competitive? How can they develop value innovation while considering ecological sustainability? If the answers are not promptly available, carmakers are at least well aware of the challenges they face. Technologies associated with low (or zero) emission vehicles can either generate or destroy value, hence relating directly to sustainability strategies. The choice has implications in almost every sphere of action, from car design to the technologies associated with systems of production, marketing, sales, distribution and maintenance, with some technologies also requiring changes in refueling infrastructure. The race is underway. Broadly, carmakers are investing in three main areas that can result in vehicles with low $CO_2$/Km: small cars, cleaner fuels and alternative powertrains. The following session briefly presents these options, their advantages and limitations.

### Smaller cars, cleaner fuels and alternative powertrains

The rationale for promoting small cars is straightforward: they are lighter[27] and consume less fuel, which means less $CO_2$ per Km. Today the Mini segment encompasses the extreme small with the *Smart ForTwo* championing the lowest levels of $CO_2$ with 95 grams per Km. For the demise of environmentalists, though, the choice for small cars had historically little to do with ecological concerns. They have either been a cost issue, since small cars tended to be the cheapest cars in the market, or an issue of status. Indeed, in the past decade or so, some carmakers tried to avoid razor-thin margins by investing in differentiation strategies for the Luxury Mini segment. By doing so, they could rely on consumer willingness to pay price premiums for some models. There are mixed results, with the revamped version of the British *Mini* as the most successful story, as well as the *Volkswagen* (VW) *Beetle* and the *Fiat Cinquecento*, which is doing well in its first years after its launch in 2006. The *Smart ForTwo* is more controversial, since it was a totally new car concept that initially aimed at being identified as a green car. However, in its early days, the eco-attributes of the *Smart* had little weight in the purchase criteria of consumers. Only when the marketing strategy refocused on its appeal as fun-to-drive, and its high-tech equipment, such as formula one soft-tip gearshifts and the design of interior parts that consumers started to buy it.[28] Although the project remains promising, after ten years of operation, the Micro Compact Car (MCC) company is still not profitable.[29] Overall, differentiation strategies in the Luxury Mini segment focus predominantly on the emotional appeal (some sense of nostalgia, in the case of the *Fiat Cinquecento*, *Mini* and the VW *Beetle*) and, therefore, are difficult to replicate. Such appeal, however, is clearly limited to relatively small market niches.

As a whole, there is no doubt that the small-car segment (cheap or expensive) will remain attractive to both consumers and automakers. After all, this is the easiest way of selling cars without changing existing systems of production and consumption. Carmakers can continue to produce all-steel cars powered by small ICEs. However, if this simplifies production, in most cases it does very little to help carmakers' finances. As the segment becomes overcrowded, profits erode. The launch of the *Nano* in 2008 is exemplary in this respect. At less than 100,000 rupees (US$3000) for the basic model, the Indian company Tata Group intends to sell 350,000 cars per annum in India. The *Nano* is following on from the success of the low-cost *Renault Logan* (sold at around US$5000) and the *Fiat Palio*. Other carmakers from emerging economies, such as Proton (Malaysia), Daewoo (Korea) are already following this market lead. Although there is space for individual successes within the segment, crowded markets amplify the well-known proverb in the industry: small cars equal small profits. As rivalry intensifies, profit margins shrink. Small and cheap cars are not exactly the solution carmakers can rely on to remain profitable.

In sum, when compared with larger vehicles, small cars are indeed greener, and constitute a simple way for carmakers to address regulatory and consumer demands without major efforts. The drawback is an increasingly crowded and bloodier market with ever-smaller profit margins. In addition, crowded can also have a literal meaning here.

With continuous growth of the car population, the absolute number of cars on the roads undermines any emission reductions per vehicle, representing a limiting factor for the small cars option in the long run.

Besides smaller cars, greener motoring can also be done via cleaner fuels. There is a wide range of cleaner fuels that can be used in traditional ICEs. Natural gas, for instance contain less toxins than diesel or gasoline and is currently used in many parts of the world, where it is available. However, if availability were not a bottleneck for its widespread use, natural gas is still a fossil fuel that generates $CO_2$, and, therefore, does not constitute a sustainable solution. Bio-fuels, on the other hand, are seen as carbon neutral, since the plants, while growing, previously absorbed the carbon emitted by burning them.[30] As it was explored in Chapter 6, today ethanol from sugarcane constitutes an economically feasible solution in areas appropriate for the crop, such as the Southeast states of Brazil, Southeast Asia and some African countries. Brazil managed to become the world leader in the use of ethanol as a substitute for gasoline. Bio-fuels in general remain an important constituent of the Brazilian fuel production, having recently stirred the increase in domestic-car ownership levels.[31] Although the long-term sustainability of the program should not be taken for granted, the production of the alcohol is becoming increasingly eco-efficient.[32]

The situation is quite different for ethanol made from corn. In most parts of the world, both the environmental and economic benefits of using such crops are questionable. In the US, for instance, intensive corn farming has historically survived due to government subsidies. The conversion rate from corn to ethanol is much lower than sugar cane, also requiring higher rates of inputs during farming and processing. Although different Life Cycle Assessment (LCA) methods generate different results, producing ethanol from corn may require more energy than what is extracted from it (for a broader description of LCA, see Chapter 5). Cellulosic ethanol, on the other hand, may yield better results but only if a technological breakthrough happens, with alcohol being extracted from cellulose with the help of enzymes at viable costs. Although such technology would allow the continuous use of ICE in a more sustainable manner, in the short term it remains only a hope.[33] Unless cellulosic ethanol becomes a reality soon, land use to grow bio-fuels competes with land available to grow food crops. As the price of agricultural commodities grows – as a result of higher oil prices (which influences the price of fertilizers and pesticides), the economic affluence of emerging economies, and the demand for food – bio-fuels will also be increasingly expensive. Substantial increases in bio-fuel production also require massive use of additional land to grow sugarcane or maize, which has stirred debate about the trade-offs in using agricultural land to produce fuel rather than food.[34] The account is simple: by 2020, we expect to be around 7.5 billion people and 1.3 billion cars on earth. With only 3 percent of the planet area available for agriculture, the trade-off is clear between feeding people and fueling cars.

A third option to address the environmental impact of cars during use, currently under consideration by carmakers, is alternative powertrains. Alternatives to the ICE to power vehicles have been available since the very invention of the automobile. After all, ICE was the winner of a race that included steam engines and electric motors for car traction.[35] But differently from most part of the 20th century, only in recent years alternative sources of power (or powertrains) were taken more seriously by carmakers. The combined effect of higher fuel prices and the effects of climate change explain such move. Although the range of possibilities is quite vast, the main technologies for powertrains converge into hybrid and pure Electric Vehicles (EVs).[36]

In the past decade, Hybrid Vehicles (HVs) became increasingly popular. Hybrids consist of the combination of two sources of power: an ICE or a fuel cell, and one or more electric motors. Using a parallel or serial system, a hybrid car operates by using the powertrains concurrently or independently.[37] Hybrid technology permits electric batteries to be charged on board via normal ICEs. The result is a car with improved fuel economy and, consequently, lower $CO_2$ per Km. The hybrid technology is not new, and since early 1990s most automakers have been presenting hybrid *concept cars*[38] in motor shows. Toyota, however, managed to be the first carmaker to produce hybrids in large scale, selling more than one million units of the *Prius* since its launch in 1997.[39] The success of the Prius caused many to believe that hybrid technology would dominate the alternative power source scenario. A closer look into HVs, however, reveals serious limitations.

Hybrids present 30 percent more components than a traditional ICE car, adding costs and weight. With today's technology, HVs are definitely not competitive in low-end segments. Indeed, the available hybrids in the market, as well as the ones in development phase target high-end consumers with luxury brands.[40] From the environmentalist perspective, things are not much brighter. Although HVs have a much improved fuel economy, fossil fuels still power the vast majority of them. Today's hybrids also use all-steel monocoque bodies, resulting in heavy vehicles that need substantial power to transport the car mass.

Even the plug-in hybrids fall under the same problem. Recharging batteries at home may further improve the range but the (heavy) vehicle still relies mostly on ICE for long-range driving. Overall, conventional hybrids (powered by ICE and electric motors in all-steel bodies) can be a satisfactory solution to increase the fuel economy of large or luxury cars. They are not economically viable solutions for the vast majority of models currently in the market.

Other alternative powertrain technology, well known to the automotive industry, is the (pure battery) EV. Although EVs have been in use since the early days of the automobile, when compared with traditional cars, their relatively low speed and limited range confined them to marginal applications. The limited range makes them ideal as a neighborhood vehicle to perform short trips in the vicinities of households. But, the lack of recharging infrastructure has historically compromised its chances to succeed in mass market applications. If bought and used as traditional vehicle (i.e., requiring high range and speed), EVs have little chance to succeed in the marketplace, as it has been the case in the past 100 years or so. In order to have a reasonable range, EVs need a substantial amount of batteries, which end up representing 30 percent of the vehicle's costs. The case of the *Think* car, discussed in Chapter 1 is exemplary in this respect. One of the reasons for the failure of the enterprise to sell high volumes of the EV was high price, when compared with conventional cars. After all, who wants to pay a price premium for a car with reduced range and speed? If used as conventional vehicle, the cost of batteries limits the chances of success in low-cost and even medium-range markets. A good example of this is the electric sports car developed by Tesla Motors. At US$110,000 per vehicle, there are still questions whether the enterprise can be profitable.[41]

An intriguing question to ask, then, is: if the technology is not satisfactory, what explains the revamped interest in EVs? The answer is simple: the technology encompassed by EVs is central for almost any direction the auto industry will take in the coming decades. Long-term alternatives for the ICE car – HVs (vehicles combining ICE or fuel cells with electric powertrains) and pure EVs – all involve electric traction.[42] The issue is whether to generate electricity on board (hybrids) or elsewhere (pure electric). There is no question, however, whether powertrains of future cars will use electric motors. This poses a great dilemma for automakers. Adopting electric traction will not only make existing engine production systems obsolete but also impact on the car body and chassis. A switch to electric traction exposes the problems of vehicle weight, which are currently obscured by the power delivered by ICEs. Overall, "any move to increase the number of EVs is likely to have a greater impact on the basic design of the motor-car (...) A significant move towards EVs has serious implications for our existing transport infrastructure as well as for the design of cars themselves".[43] No wonder the historical schizophrenia of automakers regarding EVs. Timid attempts to promote them have often served the argument that EV technology is not viable. Will this change with a new reality of high oil prices and pressure to curb carbon emissions?

## New technologies, undifferentiated markets

Should carmakers expect any enduring competitive advantage emerging from smaller cars and alternative powertrains? Even better, can such technologies help automakers to create new market spaces? As discussed in the previous section, smaller cars, cleaner fuels and alternative powertrains are the available options for automakers to render their vehicles low emissions. Starting with cleaner fuels, let us briefly examine the case of ethanol, used by 90.6 percent of the cars sold in Brazil today (see Chapter 6 for more details on the Brazilian ethanol program). What has been the main advantage resulting from selling ethanol-powered cars? The answer is a clear public benefit of having a renewable fuel source. If bringing such products to the market resulted in any private profits in the form of an early mover advantage for automakers, it was a short-lived one. The switch to ethanol vehicles had limited influence over both the business model of carmakers and their market share. The technology to power ICEs with ethanol quickly became omnipresent in the country. Today, the majority of carmakers operating in Brazil offer the flex-fuel option (gasoline and ethanol) in almost every model. The lesson is clear: Cleaner fuels for ICEs do not generate lasting competitive advantages for vehicle manufacturers. The reason is even simpler: The flex-fuel technology did not change the basis of competition, which remained focused on specific strategic groups. Any type of fuel used in ICEs (clean or otherwise) requires economies of scale to be profitable for fuel retailers. As a result, vehicle suppliers have a strong motivation to share the technology for a specific fuel, rendering it a non-competitive issue, as discussed in Chapter 1.

What about smaller cars and alternative powertrains? Again, should carmakers expect any enduring competitive advantage or new market

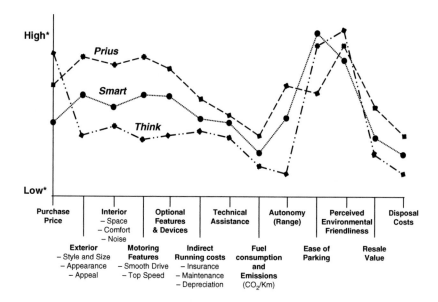

FIGURE 7.1 **Strategy Canvas of the Smart, Prius and Think**
\* Values for *high* and *low* are not commensurable across the chart, intending only to contrast the three
value curves. The curves were inferred by the author.

spaces emerging from them? Figure 7.1 presents three value curves,
which are associated with low-emission vehicles. Observe the curves
of the *Smart ForTwo* (a small car powered by an ICE), and two different alternative powertrains: the *Prius* (a hybrid car) and the *Think* (an
electric car). In the case of the *Smart ForTwo*, which was launched in
the same year of the *Prius* (1997), being the world champion in low
emissions (95 g/km) also means low consumption and, as the curve
of the *Smart* in Figure 7.1 indicates, the extremely compact size and
some motoring attributes appeal to a selected group of enthusiasts.
However, its main advantage has been the reduced size, which makes
parking easy in crowded European city centers. Indeed, size explains
particularly well the early success of *Smart* in the Italian market, where
parking is a chronic problem.[44] However, as similar models enter the
market, as it has been the case of Fiat *Cinquecento,* such appeal may not
be sufficient to keep clients faithful to the *Smart.*

When it comes to the *Prius,* the vehicle indeed opened the market
space for mass commercialization of HVs, and resulted in a remarkable spin-off value for Toyota. The media attention gained by the *Prius*
transformed Toyota into the sustainability leader of the industry. If
measured in reputational value, the *Prius* certainly represents a success
for Toyota. In fact, considering both the high cost of manufacturing

the *Prius* and its relatively high market price, it is only when the reputational value is included that the *Prius* represents a value innovation for Toyota. And, even so, the *Prius* certainly did not create Sustainable Value Innovation (SVI). The reason is simple: as mentioned earlier, hybrid cars reduce the overall impact of motoring only marginally. Besides the credits of first-mover advantage, the success of the *Prius* has more to do with the fuel economy than any other eco-oriented attribute. Essentially, the value curve of *Prius* in Figure 7.1 shows that it is an expensive all-steel sedan powered by an ICE, helped by electric motors. As competitors run to develop their own hybrids, Toyota's advantage with the *Prius* will eventually erode.

Finally, observe the case of the *Think*, the Norwegian EV briefly discussed in Chapter 1. Figure 7.1 shows that, when compared with both ICE cars (*Smart*) and hybrids (*Prius*), the *Think* presents inferior attributes. As long as the *Think* (and, for this matter any other EV) is marketed as a conventional vehicle, its high eco-attributes will always be counterbalanced by the key motoring attributes of conventional cars, such as top speed, range and zero-to-100 km response. The result is what everybody knows: market failure. Breakthroughs in battery technology and widespread user-friendly recharging infrastructure may change this (as it will be discussed later in the chapter). However, the key argument here is that the technology *per se*, being hybrid (*Prius*), pure battery EV (*Think*), or small ICE engines (*Smart*) does not generate lasting differentiation for automakers. Although some powertrain technologies may bring a first-mover advantage, as it was the case of *Prius*, it is a short-lived one. Over time, they tend to converge to a strategic group within a core segment for cars, resulting in undifferentiated markets. If "strategy is being better by being different",[45] carmakers certainly have a problem here. In order to solve it, they need to consider not only the technologies embedded in cars but also the business models to deliver them.

## Time for a u-turn: from car models to business models

The auto industry is under increasing pressure to change. In terms of rivalry and economics, it is immersed in a *red ocean* of price wars of ever-decreasing margins. Competition from emerging economies – India and China in particular – is making red oceans even bloodier. As the iconic launch of the *Nano* car in 2008 by the Indian Tata Group suggests, rather then being passive markets for Western carmakers, local producers will not only increase rivalry in their emerging economies but also become new players in established markets.

In the past decades, a disjunction between what consumers want and what automakers are able to offer, profitably, has grown substantially. While consumers have been looking for individuality, innovation, customization, aesthetic quality measures, differentiation and product-in-use services, carmakers tend to focus on stability, uniformity, capacity utilization, automation, capital equipment, tolerance quality measures and, ultimately, economies of scale. The costly system of building cars for stock, rather to order, generates huge inefficiencies throughout the value chain and, more worrisome, customer dissatisfaction. The result is a system in which both sides are discontent: carmakers struggle to remain profitable while a wide range of customers do not get what they want – at least in the time they want. The truth is that large economies of scale have not been working for most car models for quite a while.

As global warming gets further into the political agenda, reducing emissions of $CO_2$ is increasingly becoming an imperative difficult to comply with. Not only technological hurdles are on the way but compliance may be quite costly for some carmakers.[46] The implementation of cleaner fuels and alternative powertrains will lower the impact per vehicle, but the pressure will not cease until manufacturers' portfolio is mostly composed by Zero Emission Vehicles (ZEV). Altogether, the *cloudy skies* will remain so for the time being. More disturbingly, and to the surprise of some, not even ZEV will help carmakers be more sustainable. Smaller cars, cleaner fuels and alternative powertrains will certainly help carmakers to reduce $CO_2$ emissions. However, they will not ease the problem of low profitability of the business. The roots of the industry's problem are to be found elsewhere: systems of production based on the concept of all-steel monocoque car bodies,[47] and systems of consumption based on All-Purpose Vehicles (APVs). The first one goes back to 1914, when Edward Budd's monocoque technology made possible the production of high volumes and facilitated automation in manufacturing of the modern automobile, thereby underpinning both the capital intensity of the automotive industry and the requirements of economies of scale. The dominance of steel in car manufacturing also resulted in vital consequences for the future of the automotive industry. For it required high investment in manufacturing technology, the production of all-steel monocoque car bodies precluded future innovations in production processes, as well as the use of alternative materials. Joining the Budd paradigm involves an entry cost ranging between €560 and €800 million, without the dedicated tooling for a particular car or model.[48] Such investments in pressing and tooling require each new model to reach break-even points between 100,000 and 150,000 cars per year to be profitable. Although some vehicles can sell high volumes, market fragmentation has been leading to lower productions runs, making success stories, such as the Renault-Dacia *Logan*,[49] increasingly scarce. Steel also results in heavy vehicles that are costly to run, since the weigh considerably reduces fuel performance. Overall, the all-steel monocoque technology limits the chances of carmakers to break-even at low volumes, while bringing additional economic and environmental costs throughout the life cycle of the vehicle.

Another overlooked problem is the high level of redundancy embedded in vehicles. Modern cars are designed to be APVs,[50] which are able to perform various roles competently but none with great efficiency. Today, most volume cars can be classified as APVs.[51] Although the legal speed limit in most countries is 110 km/h and the average traffic runs at approximately 70 km/h, most APVs can reach speeds of more than 200 km/h, for a range of 400 to 500 km. As most of us know, the vast majority of trips do not demand such performance. The average drive in cities – the place where most cars spend the largest part of their time – requires less than 20 percent of such performance, and the average occupancy (1.2 people per car) is also much lower than the capacity of these cars to accommodate five people.[52] Cars, therefore, embody a high degree of redundancy in design, a feature that results in both economic and environmental costs for the user. Overall, there is a high degree of inefficiency embedded in both the systems of production and consumption of the modern automobile. The technologies associated with the modern car are economically inefficient at low volumes, and eco-inefficient at any.

In addition, even if all modern cars could become low or zero emissions, we can only wonder how on earth the planet can cope with the production of 100 million new cars every year – the industry forecast for the year 2020. In a planet affected by climate change, no matter how green new cars can be, an ever-growing car population simply cannot be sustained indefinitely – from both ecological and public management perspectives.[53] Ecological sustainability requires that, at certain point, the global number of cars stabilizes and eventually starts declining. From the public administration view, as we become 7.5 billion people by 2020,[54] urban congestion is expected to reach levels that might motivate regulators to ban the private car altogether from large portions of cities. Hence, the ever-crescent car population can generate another self-inflicted wound on the automotive industry; it will ultimately limit the options of private ownership and seal the fate of some automakers.

Population concerns bring us to another sticky issue faced by the industry: its role in reducing social inequality. Considering the economic importance of the sector, much more will be expected from automakers in addressing social development issues. As poverty reduction becomes a leading component of CSR, the industry has been increasingly requested to provide economic value – in the form of goods or services rather than philanthropy – to less-privileged economic classes. Even though most people in developed countries can afford at least small and cheap models, cars have historically privileged the middle and upper classes. In emerging economies, this divide is even more evident. The installation of car factories in countries such as India and Brazil hardly serves the interest of the poor, who can only dream about individual motorized mobility. As societies become more aware of the responsibility of business in bringing prosperity for deprived social classes, in the coming years the pressure will increase for the auto industry to find solutions for sustainable mobility not just for those who can afford cars but also for underprivileged non-customers.

Altogether, one does not need to be an auto expert to perceive the disjunction between socio-economic and technological trends and the business model currently adopted by auto assemblers: revenues based on selling inefficient and polluting vehicles, which are used inefficiently by a minority. If a truly sustainable auto industry is to emerge, the logic of wealth creation needs to be seriously rethought, urgently. The task is more than daunting. Automakers will need to respond to the demand to create both private profits and public benefits, simultaneously. They will need to generate long-term value to shareholders and customers (private profits), as well as to the society as a whole, in an ecologically sustainable manner (public benefits). In other words, more than any other industrial sector, they are required to create SVI.

Undeniably, there is still a vast scope for the commercialization of cars as we know today. For diverse reasons, individual-car ownership will remain a dream for many of us. Many people will always see cars as the ultimate expression of their individuality, and having a *Ferrari*, a *Mercedes* or a big *Jeep* will continue to make them happy. As symbols of self-expression, cars will always be part of the glamour of life. The car industry can certainly make them in different ways, using more efficient materials and powertrains, and many car models can indeed be profitable and, depending on the circumstances and technology, even environmentally sustainable. Although the industry has problems, obviously not all is wrong with the existing car system. There is certainly scope for the production of high volumes of some models to specific market segments. The recent success story of the Renault-Dacia *Logan* is exemplary in showing that some car models can surely be very profitable. When the markets can absorb high volumes of the same vehicle, the old all-steel monocoque technology works very well indeed.

Nonetheless, for the industry to align itself to the emerging trends and succeed economically, diversity will be increasingly crucial. And diversity here does not refer only to vehicle sizes, sources of fuel, powertrains and body materials, but also – and more importantly – to different value propositions and business models.[55] In order to respond to market demands, car manufacturers may need to use different design concepts and materials so they can move away from the all-steel car monocoque, which imposes extremely high investments and, consequently, break-even per model. The *Think* car, mentioned in Chapter 1, showed that by avoiding the main paradigm of production used by the industry and instead using aluminum in the body structure and plastic composites for panels, it is possible to reduce the break-even point from the industry average of 100,000 to just 5000 vehicles. Although such option certainly entails its own problems, it represents the practical viability of bypassing the imperative of large-scale production. In fact, Morgan and Lotus are examples of long-living companies in the automotive sector that avoided the capital-intensive model. The key to alternative structures used by these firms is the use of technologies that require much lower capital investments. In terms of material, steel, aluminum and plastic composites have been more common in the sector, but more recently carbon fiber is making its way to mainstream applications. Carbon fiber was originally developed for the aerospace industry and later transferred to the *Formula One* sector, which could afford to retain those high costs. Today, however, it is possible to process carbon fiber in a low-tech, low-cost manner, as some small firms such as Axon Automotive currently do.[56]

The urgently necessary changes in the car industry do not render the existing car-making model obsolete, but it is necessary perfecting the automobile system by adapting it the emerging socio-economic, regulatory and technological trends. Deploying low-emission technologies and redesigning systems of production will certainly help a more sustainable industry to emerge out of new car concepts, materials and associated technologies. However, SVI can only be created if the patterns of utilization are also addressed. The reason is simple: the inefficiencies in car use are even greater than the ones embedded in the vehicles themselves, as well as in their systems of production. During use, the vast majority of cars remain idle for over 90 percent of the time and, when running, their low-occupancy rate (of 1.2 persons per vehicle) means they are severely underutilized. In management science terms – logistics in particular – such inefficiencies in car utilization represent a great potential for perfecting the system. In business language, the embedded inefficiencies in the car system means gigantic overlooked market opportunities.

## SUSTAINABLE VALUE INNOVATION IN MOBILITY:
### *GREENER SYSTEMS*

La Rochelle is an historic city in the West cost France, well known for its ecological credentials. In particular, La Rochelle has a long tradition in experiments about EVs that dates back to the 1970s. The Liselec EV scheme is a relatively recent program still operational in La Rochelle, which started in 1997. The Liselec fleet consists of 50 cars (*Peugeot 106*, *Citroen Saxo*, *Berlingo* and Gem converted into EVs) distributed in seven stations throughout the city. In order to drive an electric car under the Liselec scheme, customers pay a small membership fee and a price per kilometer driven. Cars are available at any time of day or night, for short trips or for an entire weekend. The cars are accessed via the membership card and a personal code, and parking space is free in certain areas of the city reserved for EVs.[57]

Liselec is an exemplary case of *station cars* (also known as city cars) that boasts high public acceptance. The La Rochelle system has been conceived to complement the public transportation system of the city. Cars are used preferentially for relatively short one-way trips to and from mass transit stations (trains, busses, ferries, etc.), as well as in areas where public transport is not served. As the Liselec suggests, municipalities normally own station cars, meaning they are *individualized public transport*. In other words, differently from mass transportation vehicles in which several persons simultaneously board a bus or a train, station cars are public property used by one or few individuals at a time.

Electric vehicles have been preferred as station cars because they are zero emissions, an important advantage for the use in city centers. However, cost is another motivation for the choice: EVs have only twenty percent of the components of a traditional ICE car, resulting in lower maintenance costs and prolonged lifespan. Costs were in fact a key motivation for rail companies to launch station car programs in the mid-1990s, which were trying to avoid the high costs of building additional parking infrastructure. Electric utilities were normally involved because they saw EVs as new consumers of electricity, and regulators supported the schemes as a means of reducing inner city traffic and air pollution. The Bay Area Station Car Demonstration project, established by the San Francisco Bay Area Rapid Transit District (BART) is a good example of such programs launched in the mid-1990s (coincidentally, 40 prototypes of the *Think* car, mentioned in Chapter 1, were used in the project during 1995 and 1998). As many other programs, the BART project ended with mixed results.[58] Among the barriers to scale up the experiments were the necessary capabilities in logistics for the management of such systems, the high purchase price of EVs and the costs to serve a relatively small number of users.

In general, what do station cars enterprises tell us? For the cynics, past failures are sufficient to be convinced that station cars simply do not work. Others may see a different picture. However marginal, such schemes indicate an *untapped market potential* situated between the extremes of car ownership and public transport, depicted in Figure 7.2. Basically, the dominant forms of terrestrial mobility are situated in Quadrants Q1 and Q3. Practically every vehicle for individual or personalized[59] mobility is privately owned (Q1), and the vast majority of vehicles for shared/collective use are public property (Q3). Car rental companies and taxis are examples of vehicles privately owned that are sequentially shared among several users (Q2). Although this is an option widely available in most urban centers, when compared with the size and scope of Q1 and Q3, they still represent a very small portion of the total number of vehicles on the roads. Hence, apart from traditional taxis and rental firms, other types of individualized motorization, such as station cars and car-sharing (Q2 and Q4) are marginal, at best.

The lighter circles in figure illustrate the scope (and today's marginal market presence) for alternative powertrain technologies within the dominant model of car ownership (Q1), as well as alternative for individual motorization, which do not depend on car ownership (Q2 and Q4). When compared with the size of traditional car ownership at the order of hundreds of millions, Car-Sharing Organizations (CSOs) and station cars represents only a small fraction of the market for cars. Mobility operators are also depicted at the intersection of the four

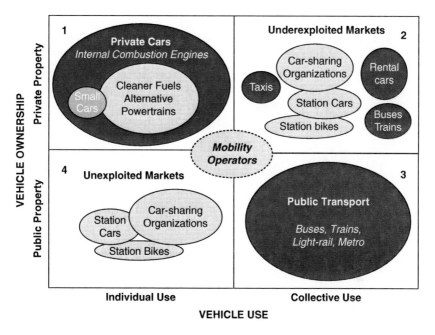

FIGURE 7.2 **Market Spaces in Terrestrial Mobility**

quadrants. These new businesses may be able to facilitate the transition to more sustainable mobility at a profit, as the case of Better Place discussed in the following section best suggests.

A brief observation of the picture may raise a simple question: What explains the marginal presence of alternative modes of individual mobility in Quadrants 2 and 4? Why carmakers have not tapped into those markets yet? Although sociologists, historians of technology and evolutionary economists have quite sophisticated answers that normally relate to system and social innovation,[60] from the management perspective, it relates to economies of scale in production, as well as the historic division between public and private domains. Regarding the first point, the previous sections in this chapter partially explained the reasons for the auto industry to have focused on the market for individual-car ownership. After all, during most part of the 20th century, huge markets were available for private cars, which expanded in tandem with the economic affluence of societies. The situation was partially reinforced by the distinctions between the role of business and governments. In democratic countries, large-scale public transport falls under the responsibility of governments, while personal motorized mobility is mainly done via private-car ownership. Even in communist societies such distinction partially holds. Car ownership has historically been an economic and/or political privilege of few. In this logic, carmakers minded their own business by supplying vehicles for private owners (including business owners, as it is the case of car fleets).

Partially, the answers also lie in the materials, technologies of production and the business model adopted by carmakers, highlighted in the previous sections. While focusing in large untapped markets for individual-car ownership, carmakers may have seen station cars and CSOs as marginal markets that were not worth the effort. After all, if they could sell cars in the order of millions of units worldwide, why should they bother with marginal volumes for CSO fleets? Even more central is the fact that the proliferation of CSOs may go against the current logic of the car business: to sell as many cars as possible. Why should makers of cars incentivize CSOs if, in the end, the success of these organizations could dent sales of new cars, resulting in lower rates of vehicle ownership? If the logic of generating income from selling cars holds, then we cannot blame carmakers for ignoring the market spaces represented by Qs 2 and 4 in Figure 7.2. They were just focusing on what made business sense – at least in the past.

In the words of Clayton Christensen,[61] this is a classic innovators dilemma: "The logical, competent decisions of management that are critical to the success of their companies are also the reasons why they lose their positions of leadership". Management decisions and concept inherited from the past impose a specific set of production technologies, significantly influencing both the profitability of the business and the eco-performance of cars. Hence, for automakers to align themselves with the emerging societal trends, it is necessary to address not only the inefficiencies embedded in the vehicles themselves and their systems of production, but also those associated with their utilization.[62] For that, automakers will need to employ a more diverse set of business models, which will eventually allow them to capture revenue streams currently underexploited. They need to move urgently into these untapped market spaces for a very practical reason: as the following cases suggest, if they do not do it, others will.

## SVI strategies within car ownership: *mobility operators*[63]

During the fall 2007, Shai Agassi, the former SAP AG whiz kid and member of the Economic Forum's Young Global Leaders, raised US$200 million in venture capital for his start-up company called Better Place, a record for a cleantech start up. In early 2008, Better Place established partnership between the State of Israel and Renault-Nissan to start the mass deployment of EVs. Israel has committed to implementing an appropriate tax policy and set out a vision by which it can become the first industrialized nation to substantially reduce the dependence on oil for mobility. Israel serves as a test-bed for applications elsewhere, mainly densely occupied city zones with short to median-driving ranges, which Agassi calls *transportation islands*. The model will develop in terms of scale to include transportation countries and transportation continents in the future. In Israel, for example, more than 90 percent of the population drive less than 70 km/day, and major urban centers are less than 150 km apart. Since EVs can run for 100 km or more without recharging, Israel presents ideal conditions for the shift from gasoline-powered cars to EVs. Better Place Israel started testing the system with 50 cars and 1000 charge spots in 2009.

Better Place has very ambitious goals. By the end of 2010 the aim is to have hundreds of EVs on the roads, supported by thousands of recharging spots and many battery-exchange stations. The year 2011 will mark the mass deployment of EVs with the diffusion of the EV infrastructure across the country. Besides Israel, in the spring 2008, Better Place set up a partnership with Denmark, a country with similarly short-driving ranges and densely populated urban areas. In October 2008, the company had signed an agreement with the Australian government and was in negotiations with several other countries.

Better Place addresses the two most critical problems for the diffusion of EVs: Infrastructure and battery costs. In terms of infrastructure blanketing Israel with battery recharging and exchange stations, will allow consumers to have coverage of EV recharging wherever they go. In a similar fashion mobile phone operators do with the installation of transmission towers, Better Place will make the recharging infrastructure available to owners of EVs, who sign up for different types of subscription packages. Customers buy mobility/kilometers as clients of mobile phones buy minutes of conversation from the phone operator. Better Place will equip every car with an on-board computer that looks like a Global Positioning System (GPS). The service control centre will help the driver to connect to charge stations, check the battery load, the availability of free parking spots with charging capacity, as well as order battery exchanges.

With battery costs at around US$11,000[64] EVs become too expensive to be fully paid upfront. Even if customers can get their money back in the form of reduced fuel costs throughout the life cycle of the car, in the past they were not willing to pay because the time to recover the investment is uncertain. To circumvent this problem, in Israel Better Place customers will be able to purchase cars from any EV dealer, and then subscribe to Better Place for service. The client will own the car, but the batteries remain a property of Better Place. Hence, the initial cost of the battery pack and the residual risk value, which has typically been considered too much of a risk by consumers, will be borne by Better Place. In order to minimize its own risks, however, Better Place establishes partnerships with battery manufacturers, such as Automotive Energy Supply Corporation (AESC) and A123. Together, they are expected to address the challenges of eventual shortages and other issues related to the lithium-ion technology.[65]

Agassi's firm presents a radically innovative business model. Better Place will also sell EVs in similar fashion telecoms sell mobile phones. The cars will be sold for much less than they cost, provided that customers sign long-term mobility service plans, which includes a chosen number of kilometers (10,000 km per year, e.g.), battery changes and replacement, and remote assistance to find parking lots, among other services. If consumers sign a long-term service contract, Agassi's company expects to provide a more affordable car, when compared to today's ICE vehicles. The basic idea is that driving an EV should be less expensive than a traditional gasoline car without compromising comfort and convenience. In other words, the value proposition of EVs should be superior to an ICE car. Early adopters, however, will have only one car model to choose from (Renaut-Nissan), before the automotive industry retools itself for the mass production of EVs.

Better Place targets the mainstream market of car drivers, the one that can transform sustainable transportation into a reality. According to Agassi, moving away from the oil dependence in individual motoring requires thinking in terms of kilometers, rather than cars, *per se*. The total kilometers driven in a year by the populations of Israel and Denmark show a very interesting market pattern. About 25 percent of cars travel more than 25,000 km/year, accounting for 66 percent of both the kilometers driven and gasoline consumed in the country (these are the frequent drivers). On the other hand, about 40 percent of the cars are driven less than 10,000 km/year, also accounting for less than 10 percent of the gasoline usage in the country (marginal users).[66] According to Better Place, these conclusions can be extrapolated to most transportation islands in industrialized countries, besides various areas in emerging economies. Hence, focusing on marginal users would do little to reduce oil consumption for personal motorization. By shifting to EVs, motorists that drive more than 25,000 km per year can make a major difference in terms of $CO_2$ reductions while representing the most lucrative client group for Better Place.

Despite a well-thought business proposition, which granted the US$200 million in venture capital, Better Place does face some challenges. A major risk is (the lack of) standardization. Today, there is a high degree of standardization for refueling any type of car powered by ICEs. Vehicles powered by Diesel, gasoline, ethanol and even natural gas can be refueled in any petrol station without hassles simply because the cars and the refueling systems are compatible. The diffusion of EVs requires similar standardization. For the system of Better Place to work optimally, the battery pack should be located in the same place in all cars, so they can be replaced quickly and safely by automated systems. Another component requiring standardization is the recharging plug. Better Place is working toward global and open standards that any car manufacturer can use.

A broader challenge is the general risk of execution, which includes the deployment and scaling up in several countries in parallel. Although the original idea was to develop and test-run the system in Israel, and only then move to other countries, the early success of Agassi's proposition rocketed Better Place to several fronts in a short time frame. The execution of the business model and the development of the operating system have been done in parallel, while Better Place extends its presence into several countries. Better Place does not insist upon government policies and regulations but is eager to work in contexts that are supportive to its mission and strategy. Such approach resulted in collaborations with various governments that support EV adoption or infrastructure build out. The governments of Israel and Denmark, for example, made the context much more favorable to EVs by instituting taxes on cars powered by ICEs.[67]

The volatility on the global financial markets after the fall of 2008 may reduce the availability of venture capital. Nonetheless Better Place remains optimistic. The company does not see the financial crisis affecting its business development in a negative way. In the coming years, governments will keep looking for viable alternatives to expensive and polluting oil for transportation, and Better Place is part the solution. Overall, the radically innovative business model and the higher prices of petrol increase the chances of the EV project to succeed. If Better Place manages to do so, it will break the long line of failures in attempting to mass commercialize EVs.

As any other business enterprise, Better Place also faces risks of failure. Nonetheless, independently on its future success, the debate provoked by Agassi's project already pointed out the central elements necessary for societies to move away from the dependency oil for individual motorization: novel value propositions and business models. Apart from any eventual difficulties Better Place may find in implementing its plan in the coming years (and they are many), the enterprise already proved that practical solutions for sustainable mobility do not require technological breakthroughs. Rather, solutions entail a systems approach. The mass commercialization of EVs has less to do with the embedded vehicle and battery technology and more with the systems supporting it. If the majority of trips for most drivers require an inferior range than EVs can offer today, the main problem is not range but rather a reliable recharging infrastructure so drives can feel confident they are not going to be left alone. After all, it is not the range of 500 km that makes people buy cars powered by petrol but rather the certainty that ubiquitous petrol stations are available for refueling. Hence, the conclusion that, before selling EVs it is necessary to have the infrastructure in place, should not surprise anyone. However, for most of the history of the auto industry, carmakers positioned themselves as suppliers of cars only, and transferred the responsibility to setting up infrastructures for EVs to governments or energy utilities. While oil was abundant and cheap, such positioning worked well simply because the fellowship with oil companies was born with the invention of the automobile.[68] If such positioning served their interest in the past, it may not suit their present needs.

Better Place suggests that, similarly to what happens today with the value chain of motor vehicles, the main revenues within an electric-oriented motorization are not going to be cashed by the makers of cars. As it was argued earlier in the chapter, the new market for EVs may not represent a solution for many of the carmakers' troubles. First, electric traction exposes the problems of vehicle weight, currently obscured by the power delivered by gasoline. Heavy all-steel EVs may be acceptable while competition is still catching up, but more than ever vehicle performance depends on the use of light materials for the structure and panels, such as aluminum and carbon fiber. Second, in the same way it happened with a large number of cell phones, the market for EV plans, such as the ones proposed by Better Place, may force the commoditization of EVs, forcing prices down. Although there is certainly scope for luxury brands to remain in their domain, powered by electricity or otherwise, the emerging market for EVs may look quite similar to current market reality of razor-thin margins in most car segments.

Overall, there is no question that Better Place is a Sustainable Value Innovation (SVI) within the dominant model of individual-car ownership. The mobility operator business plan increases the value for consumers at lower costs (private profits), while reducing the overall emissions of $CO_2$ in transportation (public benefits). By applying for a 4–6 years mobility plan, customers will pay less for the car upfront. During use, they will spend the same amount or less per kilometer using electricity than they currently spend for gasoline. Using the logic adopted by the company, a reduction of the total amount of $CO_2$ emission in mobility requires that frequent drives convert to EVs. They represent more than 60 percent of the total emissions of the sector and, therefore, are central pieces in the sustainable mobility puzzle. They are not alone, however. The research conducted by Better Place showed another interesting facet: 40 percent of car owners drive less than 10,000 km per year. Considering the break-even point of car ownership is between 15,000 and 18,000 km/year,[69] this means that 40 percent of the drivers are simply overpaying individual mobility. This information suggests that alternative modes of individual motorization may address such low cost-benefit ratio. More importantly, here is where one of the classic examples of Product-Service Systems (PSS), discussed in Chapter 6, becomes central not only to reduce the environmental impact of motorization but eventually to solve (some of) the economic problems of some carmakers.

## SVI strategies beyond car ownership: *mobility service systems*

Mobility, a Swiss company, is the most enduring success of Car-Sharing Organization (CSO) in the world.[70] The company was created by the amalgamation of two[71] small car-sharing cooperatives founded in 1987 in the German-speaking areas of Switzerland. Cost savings and environmental protection were the main motivations of the few neighbors who decided to found the cooperatives. They could hardly imagine that 20 years later the organization would count with 80,000 members and a fleet of 2000 cars distributed across 1050 stations in Switzerland.

Car-sharing should not be confounded with car-pooling, in which the vehicle is simultaneously shared by a few people. Clients of a CSO, on the other hand, use the vehicle in turns, sequentially. Customers of Mobility, for instance, can choose among a wide range of vehicles, from *Fiat Panda* to *Alfa Romeo* and minivans, to use the cars on their own, according to their needs. Members pay a monthly fee and a price to use the vehicles, which includes fuel and parking, in case it is not a station owned by Mobility. The fee includes maintenance and insurance for the cars and passengers. Using the system is quite simple: Members book the car via the internet or telephone and pick it up at the parking station where the car should be available at the required time. All vehicles are equipped with an on-board computer, which opens the doors when the membership card is swiped through a spot in the front screen. Once the car is open, the user accesses the keys and goes for the drive,[72] which is normally a round trip (i.e., back to the same parking station). Easy access to vehicles, which are normally parked close to train stations, airports and residential areas certainly contributed to success, but Mobility also has a 24/7 customer service that helps customers to find the car they want, where they need, on any day of the year.

Mobility is not an isolated case. There are a large number of smaller CSOs distributed across most EU countries and Asia. Curiously though, the world's largest car-sharing firm is a much younger American enterprise. Zipcar was founded in 1999 by two Americans who got the idea while vacationing in Berlin. In less than ten years, Zipcar has grown to 250,000 members in 50 locations in the US, Canada and the United Kingdom (UK) owning more than 5000 vehicles. Interestingly, most of this growth occurred quite recently, between 2006 and 2008. Zipcar operates in a similar fashion Mobility does. Members pay a yearly or monthly fee, as well as the time for using the car. But differently from Mobility, the fuel, parking and the car insurance are included in Zipcar's price, which depends on location. In cities like New York, prices are higher than in the countryside. Similar to rental car companies, car-sharing customers choose the car they would like to drive. The Zipcar selection includes *Cooper Minis*, *BMW*s and pick-ups. Possibly, this is one of the reasons for 30 percent of the Zipcar members to have sold their own vehicle, or held off buying a new one after they became members.[73] When compared with private car ownership, car-sharing services give the customers the possibility of owning several cars at reduced costs. Such value proposition may increasingly appeal to specific segments of car owners, as well as people who do not have a car.

Cynics may wonder: If there are economic and environmental benefits of car-sharing, why do these schemes remain marginal?

Why carmakers have historically neglected such market spaces? Some emerging trends should be sufficient to explain why CSOs have a better chance now. They are the trends toward environmentalism in a carbon-constrained world, addressed in Chapters 3 and 4; higher oil prices, or the approaching of peak-oil,[74] and; Corporate Social Responsibility (CSR).[75] According to Kim and Mauborgne trends are crucial in the process of creating Blue Oceans (Path 6: look across time)[76] because they are irreversible, have a clear trajectory, and therefore are decisive to the (automobile) business. Besides the importance of observing trends, there are five additional paths, which managers can follow in order to reconstruct market boundaries. Since these paths can also indicate potential value innovation for the auto industry, they may also be the first steps toward the identification of SVI strategies. Hence, throughout this and the following sections, these Paths are briefly mentioned (within brackets), to suggests potential to create new market spaces in the domain of individual mobility.

Car-sharing systems fill a gap between taxis, which are normally used for short trips and require less than one hour, and rental cars, which are normally rented for periods of 1 to 7 days (See Q2 in Figure 7.2). While jumping into a taxi involves no bureaucracy at all, renting a car entails to fill out paperwork and to get the car from the rental firm and later return it. Besides, differently from taxis, which pick you up, we normally have to go to the few parking lots of car rental companies to access the vehicle. These burdens motivate users to rent cars for a minimum of a couple of days, rather than a few hours. Hence, for those who need individualized transport for longer than one hour (taxi) and shorter than one or two days (car rental), station cars and car-sharing are the most cost-efficient options – if they were available. The problem is that, today, station cars and CSOs are so marginal, when compared with the size of the car ownership market, that very few of us even know they exist. They do, in many parts of the world. And if trends mean anything, car-sharing will eventually move from marginality to mainstream. Between 1998 and 2006, car-sharing has grown at an exponential rate, reaching 350,000 members worldwide in 2006.[77] Although the figures are comparatively small, when confronted with private car ownership, they are certainly not irrelevant as a market tendency (observe Blue Ocean Path 1: Look across alternative industries. Car-sharing can be an alternative industry for carmakers).

Those who had a chance to access car-sharing schemes, have a few common features: they appreciate the transparency of travel costs and their contribution to limiting traffic congestions.[78] For previously non-car users, car-sharing membership certainly adds to total mileage of cars, but the increase is only a fraction of the 40 percent or so reduction of mileage of car owners who join a CSO. When considering only the costs of owning a car, the potential for car-sharing is remarkably high. Some studies[79] put the break-even for car ownership between 15,000 and 18,000 km/year, which means around 70 percent of car owners would save money by shifting to car-sharing. Even more conspicuous are the savings for about 40 percent of cars users who drive less than 10,000 km/year.[80] However, this assumption does not consider non-financial benefits of car-ownership, such as prestige, identity enhancing and immediate car accessibility. When factoring in non-financial aspects, the potential for car-sharing decreases substantially, to around 10 percent. Such number, however, needs to be put in context. Emotional and functional aspects of car-sharing may change as CSOs adopt supporting information technologies and become more convenient. More importantly, branding exercises my reinforce the membership status and counterbalance the symbolic effect traditionally attached to car ownership.[81] (observe Blue Ocean Path 5: Look across functional and emotional appeal of buyers. At least part of the emotional appeal attached to private-car ownership can shift to CSOs).

When confronting the trends toward low-carbon emission, high petrol prices and addressing the needs of underprivileged non-customers, it is clear that economic viability of many carmakers will be increasingly under jeopardy. Creating wealth while addressing the demands for a better world means only one thing for the car industry: a gradual move from products toward services; from revenues based on selling cars to the ones based on mobility services. From whatever angle we look at, the gradual servicization of the car industry seems to be the only long-term solution; CSOs may represent the best chance for carmakers to start a mid-course correction. By moving into this (relatively) untapped market space, carmakers will be able to address consumer, environmental and regulatory demands, simultaneously. In other words, they can create SVI.

Consider the market spaces for CSOs in highly industrialized countries. Even if only 10 percent of the current car owners were willing to become members of a CSO, as some studies mentioned,[82] the global market is still in the order of millions of clients. The early adopters of a CSO may be composed by current car owners who are in the low-end market; those to whom the costs of car ownership is considerably high. In emerging economies, such scenario is even more crucial, since motorized transportation is a luxury that many car owners can barely afford. These people have enough money to own their cars, but just. They are potential shifters susceptible to a new value proposition (observe Blue Ocean Path 2: look across strategic groups within industries. As CSOs become more professional and convenient, a portion of current cars users in low-end segments may shift to CSOs).

When considering non-customers, who can only aspire to access individualized transportation, the number may move to the order of hundreds of millions.[83] In a world where the only options for motorized mobility are collective transport, which can be crowded dirty and smelly, or the comfort of driving a private car, there is no need to ask why people want to be car owners. Access to individualized motoring does not automatically mean, however, car ownership. It is only because the choices available to most consumers are restricted to the extremes – individual car ownership or collective public transport (Qs 1 and 3 in Figure 7.2) – that they opt for one of them. Car ownership is only one form of individualized mobility. At least for current car owners in the low-end segments and for users of public transport who are close to affording private cars (non-customers), car-sharing schemes and station cars may satisfy their needs at lower cost than private ownership (observe the Blue Ocean logic of reaching beyond existing demand, by understanding non-customers needs[84]).

The empirical existence of a large number of relatively small CSOs makes the business case for such underexplored markets. Although the scale of CSOs is still minor when compared with private-car ownership, the embedded value proposition of car-sharing constitutes an empirical example of SVI within the mobility system. Compared with car ownership, CSOs provide cheaper mobility at both lower costs and environmental impact. Current CSO membership indicates that car users require only around five percent of the cars that would be necessary if they were owners. Besides, car-sharing motivates changes in driving behavior, reducing mileage per person. More importantly, when compared with individual car ownership, the value curve of car-sharing is clearly differentiated. Figure 7.3 depicts the strategy canvasses of car-sharing in relation to the *Prius*, the hybrid sedan of Toyota, mentioned previously, as well as the *Nano*, the cheapest car in the market today, lunched by the Indian Tata Group in early 2008.

Car-sharing membership eliminates the purchase price of cars and reduces the costs associated with car ownership, such as insurance, maintenance and depreciation. In times of credit difficulties such as

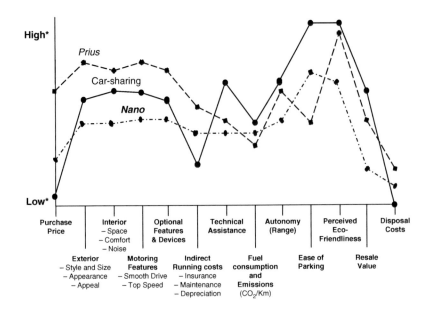

FIGURE 7.3 **Strategy Canvas of Car-Sharing**

* Values for *high* and *low* are not commensurable across the chart, intending only to contrast the three value curves. The curves were inferred by the author.

the one of 2008–2009, the elimination of the purchase costs is certainly appealing to a large segment of consumers that rely on credit for the purchase of cars. The substantial reduction of ownership costs is certainly an additional advantage. On the other hand, because sedans mostly compose the fleets of existing CSOs, it represents a notable *rise* of the car features, when compared with the small cars, best represented by the *Nano*, which tend to present inferior features. Obviously, such feature represents only an average because a large number of car models form the CSOs fleets. More absolute, however, is the rise in the value of technical assistance (provide by the CSO at any time), ease of parking (CSOs have multiple parking lots across cities) and the perceived eco-friendliness of driving (there is a reduction in the overall driving mileage per person). Overall, when compared with both the strategy canvasses of *Prius* and *Nano*, car-sharing presents a clearly differentiated curve. However, while the value proposition of a cost-driven CSO may appeal to many current car owners in the cheaper segments represented by the *Nano*, as well as current non-customers of cars (mainly in emerging economies), it is certainly not the case for owners of luxury brands and niche segments. Hence, it is fundamental to recognize that car-sharing is not for everyone; it works for selected segments of the market, currently underexplored.

Since small CSOs operate not only in several European countries and in the US but have also been popular in Singapore, Japan and China, the business case has already been proven. Their existence also means that they are not a totally new market space. Nonetheless, when compared with the gigantic size of private-car ownership, the market for CSOs is marginal at best. In addition, the marginal size of existing CSOs, as well as the limited resources available to most of them suggests that, from the point of view of carmakers, this can be considered a new market space. Hence, the argument here is less about the viability or novelty of CSOs and more about the potential to scale them up, to a level that would motivate the auto industry to be a major player in this domain. Indeed, the limited scope of most CSOs has been their main weakness, putting off many potential customers. It is easy to understand why. If you are buying the convenience of a system, you do not accept a half-ready one; either the software you bought for your computer works or it does not; either the car you need is easily accessible (the parking lot is close enough to your house or someone brings the car to you), or it will not satisfy your needs. A quasi-ready system with a few stations and clumsy bookings and control systems is not ready for commerce.

An example of a well-planned and well-managed mobility system is the Velib *station bicycles*, launched in July of 2007 in Paris; by December of the same year 1451 stations and 20,600 bikes were available to the public. The numbers speak by themselves; in its first year, 25 million trips were registered. Velib was an instant success because the system worked closed to its full potential at day one.[85] Since the stations are not more than 300 meters apart, potential users in Paris can find a bicycle easier than they find a metro or taxi station. Velib's instant success is significant because it breaks a long tradition of failures in implementing similar schemes in European cities. More importantly, it indicates that the convergence of Information Technology (IT), environmentalism and public–private partnerships have the potential to unleash social innovation. For the Velib users, sharing a bike for ride is not an extravagant behavior of a greenie hippie but just another way of moving around the city. The success with bikes motivated public administrators of the city of Paris to extend the system to EVs.[86]

Velib serves as a proxy for station cars and CSOs. Think about it. Confronted with the option of joining a CSO, you either have the convenience of booking a car via the internet or phone, picking it up and returning it without any hassles, or you will rather stick to private-car ownership – even if car ownership comes with costs and hassles you would rather avoid, such as insurance, maintenance, depreciation, cleaning, etc. This means that, in order to make it big, a CSO needs to start big. In cities like London, Moscow or Los Angeles, a CSO fleet may have to be in the order of hundreds of thousands of cars at the very early stages. Clients should be able to have easy access to a wide range of cars, from a small two-seat EV, to large SUVs, luxury cars and small trucks, at any time they want. In other words, vehicles adapted to the job. From Monday to Friday, for instance, an EV to go and return from work may be just right. On Wednesday evening, maybe an ethanol-powered *Mercedes* is the most appropriate car to go for a dinner; a hybrid SUV may be the best choice for the family to go away in the weekend. In other words, the system has to provide more convenience and value than a private car, at lower cost.

Obviously, setting up such a complex organization is far from being an easy task. This difficult and risky business requires not only considerable investments but also skills in logistics, car maintenance and servicing. Large scale CSO is not for amateurs. That is why, if considered seriously, car-sharing may be the right business for carmakers to take on-board. After all, who else knows more about cars and their associated technologies? Besides exploring the embedded potential for SVI in car-sharing, by tapping into this new market space carmakers may limit the emerging defiance of outsiders. As Scott Griffith, the CEO of Zipcar, the successful American car-sharing company states: "Zipcar is more of a competitor to the automakers than to the rental car companies because it creates a lifestyle change that eliminates the need for many city dwellers to have their own car". The message seems clear enough, but only time will tell whether automakers will respond[87] to the emerging threat of CSOs with just another car model, or with a new business model.

## SVI as strategy: the business of ultimate services

In order to respond to the emerging demands and trends for a better world, vehicle manufacturers have a wide range of strategic choices. Besides being makers of cars, they can, for instance, position themselves as manufactures of propulsion systems or even as mobility integrators. In principle, there is nothing forcing them to remain in the business of selling bashed and welded metal. Neither they have to lock themselves into a specific technological option, as it has been the case of Edward Budd's monocoque concept and the ICE-based powertrain, mentioned previously. If strategy implies choices, there is no shortage of them in the business of individualized motoring. Automakers do have the option to respond to consumer and regulatory demands with more than lobbying against strict $CO_2$ regulations. In a world of peak-oil and carbon-constraints, there is no need for major technological breakthroughs for the auto industry to become more economically and environmentally sustainable. What is necessary, however, is a bold strategic move toward the creation of SVI. A SVI strategy requires the industry to move away from crowded markets by focusing on the *ultimate service* supplied by cars: personalized mobility.

Business models can be established on a new revenue basis: by becoming suppliers of Mobility Service Systems (MSSs), automakers can generate revenues alongside the use-phase of vehicles – indeed the most profitable segment of the entire value chain. For instance, by supplying vehicles for its own car-sharing fleet, a MSS provider can create long-term revenue streams via the membership and use of cars, as well as a higher level of system integration and, consequently, entry barriers. The environmental benefits associated with such move are also great. The move toward servicization in the car business allows decoupling the increased access to cars from the growth of car population. Depending on location and the profile of clients, a CSO may need only between 5 percent and 25 percent of the cars that would normally be necessary for private ownership.

Although counter-intuitive, such scenario can also have positive implications for a carmaker-turned-MSS-supplier. By the very nature of a CSO, the fleet needs to include vehicles of different sizes, with the majority being adapted to large urban centers. This implies that small low or zero emission cars will counterbalance the emissions of larger luxury vehicles. In such logic, it would be much easier for a CSO to comply with a regulation requiring average fleet performance (such as the one proposed by the European Commission, of 130 g/km) than a traditional maker of cars. A CSO provider has much more flexibility to adapt its fleet to the requirements of both regulators and consumers. Indeed, the fleet can be customized to current non-consumers of cars in emerging markets, in which a large number or users could share small, cheaper and low-emission cars. For instance, based on current rates of car-sharing membership,[88] a CSO fleet formed by, say 50,000 *Nano* cars (plus other vehicles) could eventually supply personalized mobility for over 1 million users in an Indian city. Instead of disputing extremely low margins of selling cheap cars, the Tata Group would generate stronger system integration around its MSS. Rather than looking for one-off revenues by selling cheap cars, Tata would guarantee long-term revenues of a much larger number of users. When considering the tripod of economic, environmental and social benefits, the commoditization of a large segment of individual motorized mobility makes more than business sense: it makes SVI.

The logic of MSS provider does not deny the emotional or aesthetic appeals attached to cars. As mentioned earlier, there is significant scope to continue selling private cars to a large number of segments. After all, as in other industry, market segmentation means that car-sharing is not for everyone. However, as Clayton Christensen emphasized: "There are a significant number of different jobs that people who purchase cars need to get done, but only a few companies have stacked out any of these jobs markets with purpose brands".[89] Car-sharing may be a way of responding to such criticism, for it allows car manufacturing to become more specialized, more profitably. To use a term coined in the 1980s, carmakers can achieve *flexible specialization*.[90] They can, for instance, reinforce the value of a particular brand by specializing on a few lucrative vehicle models, adapted to the ultimate service it is supposed to deliver. Additionally, car-sharing can become one of the various MSS supplied by carmakers. As MSS providers, they will have the flexibility to include any type of cars in the fleet, including the ones supplied by competitors, so the value proposition for consumers is as high as possible.

## WHEN *SUSTAINABLE VALUE INNOVATION* PAYS

Considering SVI strategy requires a rupture from traditional industry practices, its deployment is less constrained by the context than by the competence firms have to innovate and create new market spaces. After all, if the SVI strategy requires substantial innovations in the model of the business, its implementation is more an issue of managerial audacity than the positioning of the company within an existing competitive space. Nonetheless, as it was the case of the previous four Corporate Environmental Strategies (CES), some specific conditions do facilitate or hinder the application of SVI strategies.

### Context

In general, industries presenting great potential for systemic efficiency improvements are the best candidates for SVI strategies. As this chapter explored in detail, the system of production and consumption of private automobiles is exemplary, but a wide range of other *functional products* may fall in this category. Some market segments of products that perform a clear function are prone to be the basis for SVI. Among them are several types of office and home appliances, such as computers, fridges, and television sets, besides other electronic and electric equipment. In some cases (still requiring detailed analysis), inefficiencies embedded in systems of production and consumption represent untapped market spaces, which can be explored with SVI strategies.

In the past decades, some market segments of a wide range of functional products have been commoditized, with consequent reduction of profit margins. In order to avoid price wars of highly competitive market spaces, some manufacturers may choose to deploy business models that focus in the delivery of the ultimate service provided by the products. For instance, a former manufacturer of air conditioning may deploy a SVI based on the supply of thermo-comfort, which may entail working with insulation systems and other technologies for ecological architecture. Such move from products to the job they should perform has the potential to reduce systemic inefficiencies and associated environmental impacts, while creating additional value to customers and long-term revenues streams to companies.

Most candidate industries for SVI strategies are indeed immersed in *bloody oceans*, in which high levels of rivalry, where low-growth rates and margins are the norm. They are also under *cloudy skies* of ever-increasing demands to reduce environmental impacts. As it was discussed in Chapter 6, in the case of home appliances and electric and electronic equipments, besides the increasing competition from emerging economies, with China on the lead, the EU Directive on Waste of Electronic and Electric Equipment (WEEE-2002/96/EC) is forcing them to take back their products without transferring the costs to consumers. The result is ever-increasing pressure to cut both costs and environmental impacts. For such firms, a move toward a Product-Service System (PSS), discussed in Chapter 6, may a viable solution to address environmental and costs issues, which may eventually result in new market spaces for them.

Even though there is a scope to deploy SVI for consumables, the scope is far more reduced than functional products. Processed food, for instance, will tend to fit either Eco-branding or E-cost strategies, respectively described in Chapters 5 and 6. Food can certainly be transformed into a service via restaurants or catering but, obviously, there is no novelty here. This is also the case for a great deal of products sold by food retailers. Indeed, a brief classification of successful Blue Ocean cases reveals the vast majority to be functional products or services.[91] Not surprisingly, as an extension of the Blue Ocean logic, SVI has better chances to be successfully deployed when it is possible to transform a product-oriented value proposition into the ultimate service it provides. Such move represents the so-called dematerialization of the economy in which economic growth can be detached from the physical need for additional resources. Somehow, this is a prerequisite for sustainability at the societal level, making SVI the strategy best aligned with global sustainability.

Finally, a large number of people simply cannot afford the functions delivered by products such as cars, computers or washing machines. Many of them are tier-one non-customers, simply because such products are still too expensive for them. Although most of these non-customers are not at the base of the social pyramid, they can still be considered underprivileged social classes, which have been neglected by businesses focusing on middle and upper social classes. For these people, the only way to afford individualized mobility is via substantially cheaper access to cars. Although companies will have to define the scope and quantify the market for each functional product, new business models can push the economic frontier downwards, allowing the system innovators to reach previously neglected customers. More significantly, the market space for such type of non-customers is far bigger than the current one formed by those who can afford products and services in developed nations. In order to identify such market opportunities and deploy SVI strategies, companies need to develop a good degree of self-criticism and competences, which allow them to align emerging trends for a better world with markets forces. Tough call, but possible for some.

## Competences

Sustainable Value Innovation (SVI) builds upon the logic of Blue Ocean Strategy (BOS). As emphasized earlier in this chapter, a key element in the process of Blue Ocean development is to reconstruct market boundaries. To do so, companies need to apply the Six Paths Framework, which have been mentioned throughout the last sections, in the light of the auto industry case. The Six Paths logic helps mangers to identify new customers and develop value innovation, which eventually allow them to reconstruct market boundaries. As an extension of this logic, in order to identify SVI, managers need to include environmental and social demands to the analysis. By following any of the Six Paths, they need to check whether lower environmental impacts and higher contributions to society may emerge.

Addressing systemic inefficiencies require a detailed and, for people within the business, indeed detached analysis of the technologies embedded in the product, as well as in its system of production and consumption. Corporate leaders need to challenge themselves by questioning the fundamentals of their business. They can start with a basic question: what is the ultimate service that needs to be delivered by our business? In order to answer this question, managers will have to put aside, at least for analytical reasons, all the complexities involved in the current business, from sunk investments, technologies and regulations, to consumer emotional desires and symbolisms. The question will oblige them to think about their competences in delivering the ultimate service as efficiently as possible, with the lowest environmental impact and the highest possible contribution to society. This is certainly not an easy task. Neither there are simple answers for the following questions, which point toward the competences firms should have to develop SVI:

- How efficient is the overall system of production and consumption embedding our products and services?
- Can we substantially reduce systemic inefficiencies, as well as their associated life cycle environmental impacts by:
  - Changing materials, fuels or power sources employed in our products?
  - Adapting our systems of production to new materials or production methods?
  - Changing the patterns of utilization of our products or services?
  - Applying new value propositions and business models?
  - Changing the way the product is sold or used?
  - Generating revenue streams in the use-phase of our products?

- What are the limits of growth of our industry? If we keep doing what we do today, is there a physical limit to growth? Can the environmental impact of our business be tolerated indefinitely?
- Who are the non-customers formed by underprivileged social classes? Who are we neglecting with our current business model and consequent costs to access our products and services?

- What is the direct contribution of our business to poverty alleviation? Is our contribution only indirect, via taxes?
- Is the overall sustainability of the business dependent on major technological breakthroughs? If they do not happen soon, what is our plan?

## CONCLUSION

In order to present the logic embedded in Sustainable Value Innovation (SVI) – the fifth and final sustainability strategy – this chapter went deep into the economic and environmental problems faced by the automobile industry, as well as potential solutions leading to sustainable mobility. The main rationale for SVI strategy builds upon the logic of Blue Ocean, and embeds it on the current demands for a better world, in which corporate social and environmental responsibility is a major trend. As the case of the auto industry suggests, drastic problems require radically innovative strategies: the ones based not only on technological innovations but also on new value propositions and business models.

The chapter indicated the need for managers to consider the limits to growth of the industry with existing technologies as real, rather than put the hopes in technological breakthroughs that may never happen. For instance, hydrogen fuel-cell technology and cellulosic ethanol may become a reality for carmakers in the long-term, but betting in such technologies seems riskier than applying a bold strategy that depends on existing technologies and managerial audacity and competence. By adopting innovative approaches to the management of the life cycle activity system of their products and services, companies can create value to existing and new customers at reduced economic *and* environmental costs. For that to happen, however, it is necessary to decouple revenues from the *materiality* of products. Overall, strategies based on SVI require the skills for *systemic innovation*, so to align business needs with the gradual dematerialization of the economy. For those who succeed, the *clear skies* will form the horizon for quite a long time.

# 8

# SUSTAINABILITY STRATEGIES AND BEYOND

Sustainability strategies are choices available to managers to align environmental and social investments with the generic strategy of the company. Such investments are similar to others in business: only a few will render private profits. In the same manner an athlete fit to win the marathon holds minor chances to win the 100 meters also, a company fit to apply for an eco-label may not be endowed to compete in the bio-polymers markets or become the leading organization in setting up a Green Club. As strategy entails choice, adequate competences require some level of specialization. Companies have good reasons to generate public benefits by going beyond compliance and reducing the environmental impacts of processes, products and services as much they can (to be fit). For some, however, the possibility of generating competitive advantage out of eco-branding, for instance, is higher than other sustainability strategy (to be the best). Hence, while addressing stakeholders' expectations, managers should not be distracted from the most appropriate strategic focus for their companies. By choosing to focus on a particular sustainability strategy based on solid business principles, they are also addressing shareholders expectations. They are investing on the areas that will generate public benefits with best chances to create competitive advantages or new market spaces for the firm.

For the majority of enterprises, the market is crowded with competitors. No matter their efforts to be unique, most of them still find themselves in markets ruled by fierce rivalry. Such companies have to do whatever they can to be good citizens while remaining competitive. This means they will have to optimize resources and invest in the areas with the best chances to pay-off, so additional and more ambitious eco-investments can be made in the near future. They can prioritize these investments by analyzing the external context and internal capabilities, and then define the most appropriate Competitive Environmental

Strategy (CES) to focus on. Others may have enough brilliance and audacity to break out from the pack. They may be able to present new value propositions via radically innovative business models and eventually create new market spaces. More, they may ultimately align themselves with the new social and environmental demands, creating Sustainable Value Innovation (SVI). By increasing consumer value while generating public benefits in the form of reduced environmental impacts and value for society, these companies are closer to the sustainability goal. This concluding chapter revisits these possibilities, encompassed in the five sustainability strategies presented previously. It also brings some final thoughts about the role of business in making sustainable development a reality.

## WHEN DOES IT PAY TO BE GREEN, *REALLY?*

In the past years, stakeholders have increasingly requested companies to be better citizens and put resources into social causes and environmental protection. On the other hand, shareholders have demanded managers to base their investments on solid grounds, so to maximize the profitability of the business. As a result, executives are left with the thorny job of translating sustainability issues into pragmatic strategies, projects and practices. They find help in a multitude of materials, including theoretical approaches, tools, techniques, schemes, standards and demonstration cases of best environmental practices. Quite often, however, prescriptions to become better citizens are done as a quasi-religious dogma, in which rewards will come as long as managers have faith on the recommendations. While some level of faith and ethical commitments toward the common good should indeed be part of management values, success in business is not built on faith alone. For executives who are still uncertain about the role sustainability may have in their business, the potential of eco-investments to generate returns needs to be based on realistic management principles, rather than on the speculative believes about win–win scenarios. Unless some evidence is brought to the table, managers will not be able to justify eco-investment to shareholders, or present the criteria they used to prioritize investments in environmental or social causes.

This book showed that, as in any other management area, the return on eco-investments is conditional. Corporate idiosyncrasies, external context, as well as multiple possibilities to create public benefits, result in eco-investments competing with other business demands and, quite often, overlooked. Consider the financial sector. Some banks are finally recognizing the importance of ecological sustainability for their business – normally in the form of loans for infrastructure development projects but this is not reflected in prices of their products; even less when disaster strikes. As the financial crisis of 2008–09 showed, when (the lack of) liquidity is the name of the game, all the rest is secondary; environmental and social issues being the least important. Because investors have finite time to worry about sustainability issues, they have to balance them against other pressing problems. In general, managerial discretion limits the attention given to less central concerns. For executives who still see sustainability as peripheral to their core business, attention to the ecological dimension is more random and less strategic than it could or should be. Regulatory and market uncertainties do not help much either. Often, they create confusion about potential rewards and further limit beyond compliance behaviors. By its turn, consumer concern for nature or for the fair treatment of people and their livelihood do exist but it is often sporadic, erratic and, when compared to the overall consumption patterns, considerably modest.

This is not to say that greening does not pay; it does, but conditions apply. After recognizing that the right question to be asked is *when it pays to be green*, rather than whether it does, answering it requires a basic understanding of the competitive nature of eco-investments. Chapter 1 suggested that, out of the vast scope for corporate environmentalism, only a few strategies and practices generate economic returns, competitive advantages or new market spaces. The scope for private profits is smaller than for those in which everyone wins by working toward public benefits simply because most eco-investments that go into products need to compete with other alternatives (green or otherwise) in the marketplace. By their turn, some eco-investments aiming at reducing the impact of organizational processes require stakeholders to attribute a value to them, which is often intangible. Taken as a whole, the scope for profiting from eco-investments is smaller than the plethora of actions available to firms to reduce impacts and be good citizens.

If the scope of win–win scenarios is smaller than some people think or wish, we still need to know when greening pays. Although a definitive answer may never be possible, the first step toward answering it requires a clear understanding about what the question comprises. First, it is necessary to have a clear definition of the type eco-investment in question (Green – what type of eco-investment is under consideration?). Among many examples of eco-investments are: novel industrial processes, Clean Development Mechanism (CDM) projects in developing countries, Green Club membership, life cycle assessment of products, subscriptions to eco-labeling schemes, eco-design practices for material recovery, biotech product development, etc. A proper evaluation of such investments needs a diverse set of criteria. Although this should be an obvious realization, very often green is treated as an all-encompassing entity, leading to unfounded discussions about its value for companies. Second, the assessment of eco-investments also requires a clear time frame, as well as the consideration of the context in which the company operates, such as the political milieu, the kind of industry in which the company operates, as well as target markets and consumers (When: what is the period and the context under consideration?). For a supplier of industrial markets, a process-oriented investment such as an Environmental Management System (EMS) may yield results in the medium or long terms. A certified EMS may guarantee a contract with a demanding client firm in the long term, but the necessary training for the implementation of procedures generates costs in the short term. Besides training, once the EMS becomes fully operational, it may also yield additional indirect benefits, which may not be easy to measure. Hence, measuring its returns entail the consideration of both quantitative and qualitative data, as well as the tangible and intangible value created by the EMS (Pays: what is the data used in the evaluation?)

Besides these criteria, in order to link eco-investments to strategy, it is necessary to have a clear idea about what sustainability strategies are and, as important, what they are not. As Chapter 2 suggested, there is a major difference between what constitutes a strategic issue (important) and a clearly defined strategy (choice). For instance, reducing internal costs and managing risk and uncertainty are undoubtedly important for firms, and the reason all firms should pursue it. But cost reduction and the management of risk and uncertainty are, ultimately, rationales for strategy, rather than strategies *per se*. Since such measures are important for the success of any business, they do not involve trade-offs and, therefore, cannot be considered strategies. If strategy requires choosing one path over another, the efficient management of costs and risks constitute, in fact, what Michael Porter calls operational effectiveness. The same applies to sustainability management. Although often subtle, there are trade-offs involved in eco-investments, and their chances to increase the competitiveness of firms or create new market spaces depends on a clear strategic choice by firms. Therefore, a clear understanding about the nature of both eco-investments and strategy is the fist step in the direction of identifying *when it pays to be green*. The next step relates the development of sustainability strategies with the principles of competitive advantage.

## COMPETITIVE ENVIRONMENTALISM

The diffusion of corporate environmentalism will not reduce business rivalry. Competition and collaboration are embedded not only in the logic of capitalism but constitute fundamental drivers of human action.[1] As companies become more socially and ecologically responsible, competition may move to a higher level but it will certainly not vanish. For most players, independently on their efforts to build barriers to entry or imitation in a specific industry, rivalry will still be entrenched. Sustainability issues are no exception. As companies become better citizens, they also compete for *being better by being good*. Although collaboration is the counterpoint of competition, they are the two sides of the coin, named commerce. As some examples explored throughout the book indicated (vehicle recycling in Europe, in particular, presented in Chapter 6), collaborative efforts serve either to instill proactive or reactive behaviors toward corporate environmentalism. Overall, even if companies strive to be in monopolist positions and hence avoid competition altogether, most will have to keep dealing with rivalry, most of the time.

Strategic alignment requires distinguishing between the areas in which eco-investments can be deployed by a company. Chapter 2 indicated that, by establishing a division between the processes (or activities) necessary for an organization to function, and the resulting products/services sold by them, a much clearer picture emerges. The second necessary distinction is the one made by Michel Porter back in the 1980s, between low cost and differentiation strategies. Hence, when making eco-investments, companies may focus either on organizational processes or on their products and services. In addition, depending on the context they operate, as well as their capabilities, companies also need to position eco-investments under low costs or differentiation strategies. Figure 2.1 (p. 30) in Chapter 2 depicted the four resulting CESs in an analytical framework.

As any representation of reality, it is important to remember that the map is not the territory. The strategies are ideal types of particular phenomena, and the typology aims at simplifying an indeed very complex reality; it should not be used a straightjacket in which the categorization of each strategy is seem as an absolute. Rather, as an ideal type, the framework helps one to think about the relationship eco-investments have with corporate strategy. One should consider that in every differentiation strategy (eco-oriented or otherwise) there is a cost component; and there is a differentiation component in every cost strategy. Although the framework presents caveats such as this one, it does serve practical purposes, as the previous chapters provided evidence. Used in an *ex-ante* basis (i.e., before deploying the strategy), it becomes a powerful tool for managers to prioritize eco-investments in the light of the principles of competitive advantage. They can ask, for instance, whether investments in processes should have a priority over the ones in products, or whether there are any grounds for the company to prioritize eco-oriented differentiation over low-cost strategies. As each chapter has delved into, in order to answer these questions, managers need to know who is prone to value each eco-investment separately. Although clients may not be interested on how the company manages its factories, it may be crucial for eco-activists. The majority of consumers may not be willing to pay price premiums for eco-branding but some embedded product eco-attributes may allow the company to thrive in cost-sensitive markets.

Similarly, the framework can also be used in an *ex-post* basis (i.e., after deploying the strategy), to evaluate the benefits resulting from each strategy. Figure 8.1 (which is an extended version of Figure 2.1, presented in Chapter 2) depicts the four CES in the light of eco-investments, discussed in the following sections, as well as the outcomes resulting from them. In general, by their very nature, cost-oriented strategies (1 and 4), are more concerned with the economic bottom line of the firm and, consequently, their evaluation is relatively simpler than differentiation strategies (2 and 3), which embody more intangibles. The focus of Eco-efficiency (1) and E-cost Leadership (4) strategies is, obviously, on bringing costs down as much as possible, while observing the need to reduce environmental impacts. Although there are some intangible aspects involved in cost-oriented strategies, because the main focus is on (objective) low costs measurers, their importance is reduced. On the other hand, eco-differentiation strategies (2 and 3) are much trickier to evaluate because they relate mostly to intangibles, which are largely associated with reputation and branding.

In certain circumstances, the management of the upstream and downstream activity systems can become essential components of a

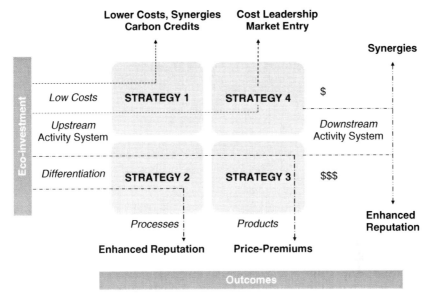

FIGURE 8.1 **Outcomes of Eco-investments in CES**

CES. Depending on how these activity systems are managed, they facilitate or hinder the efficacy of low costs or differentiation strategies. This is one of the rationales for CES to be consistent with the generic strategy of the company. For instance, Wall Mart started requiring greener practices from it suppliers, but such practices have to be aligned with the generic low-cost focus of Wall Mart. On the other hand, the endorsement Green Clubs principles (Strategy 2) or eco-labeling schemes (Strategy 3) normally increase the cost of upstream practices for companies. Returns on these eco-investments come in the form of enhanced reputation or price premiums, which are associated with differentiation strategies. Finally, firms can also make eco-investments in downstream practices. By investing in activities that eventually facilitate a close-loop recycling system, they minimize the risks of being taken by surprise, in case a take-back regulation is imposed on them. Such practices may create synergies for product recovery and recycling, as well as affect reputation positively. However, as Chapter 1 and 6 discussed in detail, the complexities involved in post-consumption management tend to transform such activities in non-rival issues, reducing the chances of generating direct competitive advantages. The following sections discuss the main outcomes of each CES separately.

## Strategy 1: Eco-efficiency

Even though resource productivity and eco-efficiency are possible in most firms, they are not uniformly distributed within and across industries. By elucidating why process-oriented companies can transform eco-efficiencies into a specific sustainability strategy, Chapter 3 mapped some of these opportunities. In general, firms that focus on Eco-efficiency strategies will benefit from lower operational costs and extra revenues via synergies, such as the transformation of by-products and waste into new businesses, as well as the generation of carbon credits. These economic advantages constitute the fundamental rationale for the deployment of Strategy 1.

Eco-efficiency strategies have greater potential to generate competitive advantage in firms supplying industrial markets, facing relatively high levels of processing costs and generation of wastes and/or by-products. Very often, because client organizations are not willing to bare costs associated with environmental protection, Strategy 1 aims at keeping operational costs down as much as possible while focusing on potential eco-efficiency sources within and around the organization. Many firms in the agribusiness and the food and beverage industries fall into this category. The potential to create synergies so to transform by-products and waste into inputs for new industrial process is greater in the agribusiness than in most industrial activities simply because organic matter is less prone to be toxic. Symbiotic relationships have indeed been part of agriculture since its early days. Such practices only changed with the advent of the modern large-scale industrial farming systems, in which monocultures are grown with the help of petrochemical fertilizers and pesticides. However, as the production of bio-fuels to power motor vehicles is pushed forward, a balanced integration of crops and industrial production seem to be more sustainable from both economic and environmental perspectives.

Eco-efficiency strategies represent a special opportunity for energy-intensive industries via several incentives for firms to generate carbon credits. Besides the mechanisms woven into the Kyoto Protocol, some governments have designed additional market-based incentives for corporations to reduce their carbon dioxide ($CO_2$) emissions. Among them are the markets to trade certificates resulting from energy efficiency measurers (White Certificates), such as the as insulation improvements in households and commercial buildings, as well as the generation of alternative forms of energy (Green Certificates), such as wind and solar power. Although future developments of these markets depend on complex and uncertain regulatory frameworks, there is a clear tendency toward the use of market mechanisms to reward eco-efficiencies. As environmental credentials became more important for the public, governments will increasingly implement these kinds of policy incentives.

## Strategy 2: Beyond Compliance Leadership

The main dividend of Strategy 2 is enhanced reputation, a primarily intangible asset. For some companies facing strong public pressure, risk management and stakeholder dialogue is the main rationale driving their sustainability strategies. Besides doing their homework to reduce environmental impacts, companies try to improve their reputation by engaging in Green Clubs as a means of legitimizing and promoting these efforts to key stakeholders and the wider public. In other words, Green Clubs are instruments for stakeholder communication and reputation management. Although Green Clubs have been instrumental in protecting companies against bad reputation, differential advantages resulting from club membership are more difficult to obtain. Since differentiation depends on reputational asymmetries, cub membership *per se* is not a guarantee of value creation. For instance, the level of diffusion of International Organization for Standardization (ISO) 14001 within a specific sector influences the importance attributed by clients and stakeholders to the ISO 14001 membership; the certification itself does not enhance the reputation of firms.

For many firms, avoiding negative reputation is already a reasonable outcome. Beyond compliance leaders, however, aim at building positive reputation via a series of programs for stakeholder dialogue, engagement and general communication. Since stakeholder engagement is built over time, such eco-investments generate return only in the medium to long term. In order to obtain differential advantages, these companies have to identify who is valuing their eco-investments. They need to identify the stakeholders who will eventually approve the company's efforts to comply with the principles of Green Clubs such as Global Compact, certify its EMS or produce sustainability reports according to the Global Reporting Initiative (GRI) guidelines. Although companies can establish effective stakeholder dialogue without becoming members of Green Clubs, it tends to work only in sectors that are not perceived by eco-activists as having high impact (independently of being true or not). Finally, companies also need to know the criteria used by stakeholders. When dealing with eco-activists, for instance, open dialogue and engagement is more effective than relying on indirect communication via reports.

## Strategy 3: Eco-branding

Similar to generic differentiation strategies, the main dividend of Strategy 3 is price premiums for eco-branded products and services. In order to obtain price premiums, companies have to rely on the willingness of consumers to pay for eco-differentiation. In the case of industrial markets (B2B), the benefits are normally translated into cost savings resulting from better performance of the product (as an input for other industrial processes), and cost reduction of risk management.[2] Although the overall willingness to pay cannot be controlled, by informing the benefits to clients, the company increases it chances of having the eco-attributes recognized by the buyer. In the B2B case, there are higher levels of rationality guiding both the evaluation of the product and the willingness to pay price premiums. Clients normally balance the extra price against the savings they make along the life cycle of the product, when compared with less efficient and eco-friendly products. In consumer markets (B2C), on the other hand, there is more room for the emotional and symbolic appeal of products. Nonetheless, because the attributes associated with the products may result in relatively lower private benefits for the consumer, they require companies to make more marketing efforts to sell the imagery of ecological responsibility. For both industrial and consumer markets, however, it is essential that the consumer is willing to pay for ecological differentiation.

In order to obtain price premiums, companies may need to invest in third party certified eco-labels, so to reduce the complexities involved in informing the eco-attributes of products and, more importantly, confer legitimacy to the claims made by them. Eco-labels can also create barriers to imitation, another essential requirement for product environmental differentiation. In markets in which most products do not present good eco-performance, an eco-label may be sufficient to keep competitors at bay, at least for a while. However, as more products acquire the same label, eco-differentiation will erode. Hence, as a rule, barriers to imitation require a protective patent of one form or another. For industrial products, holding an exclusive rights certificate will reduce the chances of imitation. In consumer products, barriers to imitation may be brought by building a robust eco-brand. As in other commercial brands, substantial marketing efforts are necessary to transform a logo into the synonymous of ecological responsibility. No matter how much information companies provide, they ultimately want consumers to trust their products. Building an eco-brand requires trust to go beyond the generic information provided by eco-labels, inducing consumers to easily recognize eco-branded products as environmental leaders.

## Strategy 4: Environmental Cost Leadership

Companies capable of exploring E-cost Leadership strategies satisfy the demands for low cost, the basic and often most important requirement of market economy. Independently on whether clients or consumers value the eco-attributes of products, such companies are able to compete in markets in which razor-thin margins are the norm. While traditional cost leadership remains the strategic focus, the intrinsic eco-attributes of products may help them to compete in existing market segments or enter new ones. Companies that are able to develop such products obviously increase their competitiveness. Besides low costs, E-cost leaders are also prepared for increasingly demanding requirements for eco-friendly products, either from regulators, consumers or both.

If this is simple in principle, it is very hard to practice. In order to reduce the intrinsic environmental impact of products while keeping costs down, companies need to stir innovation by investing in product redevelopment, eco-design, as well as new ways of commercializing them. Eco-design, for instance, may help packaging manufacturers to reduce the environmental load of their products and cope with regulatory requirements, while eventually contributing to reduce costs. In addition, by changing the nature of raw materials or entire products (polymers made out of plants, rather than petrochemicals, for instance), some companies can maintain their presence in traditional markets or access fast-growing eco-oriented ones, such as the market for bio-fuels.

As some empirical examples suggest, changing the nature of transactions between producers and consumers can also result in both lower product costs and environmental impacts. Once companies move to selling the functions products are supposed to deliver instead of selling products themselves, the interest of suppliers and buyers may converge. Rather than selling heat, suppliers may sell thermal comfort; mobility instead of cars; clean clothes instead of washing machines. In these cases, the reduction of consumption serves the interest of both parties. This is ideal in theory. Once again, while the concept of Product-Service Systems (PSS) is promising, managerial discretion and symbolic elements associated with product ownership are among the reasons for the scarcity of empirical applications. For PSS to work, systemic changes involving infrastructure and collaborations among a wide range of players are also necessary. This may be possible only by applying a sustainability strategy that aims at going beyond competition.

## BEYOND COMPETITION

Broadly, PSS define the borders between the four Competitive Environmental Strategies (CES) and the Sustainable Value Innovation (SVI) strategy, revisited in the next section. This is because PSS imply the adoption of a new value proposition and, very often, a new business model for the move from products to services. Such move can be done in both existing and new markets. For instance, a firm selling Chemical Management Services (CMS, described in Chapter 6) aims at reducing both the costs and environmental impacts of its products as much as possible. However, such CMS supplier will still be working within an existing, clearly defined market space (chemicals for factories). SVI strategies, on the other hand, have the main aim of creating higher value for customers and society in general at lower costs and environmental impacts, eventually leading to new market spaces. Since in a new market space there is no product or service to compare with, the focus moves from price to the higher value (innovation) products/services provided to customers. This is the main the logic of Blue Ocean Strategy (BOS). Rather than trying to gain competitive advantages in existing markets, companies should avoid competition by focusing on the unfulfilled demands of existing or new customers. Why should they fight for existing markets if they can create new ones? This is certainly true for players that have enough innovative capabilities, resources and, very often, the guts to deploy risky strategies. The problem is that this is not for everyone. Although most executives would rather innovate and put their companies into a monopolistic position, in practice this is extremely difficult.

If most companies operate in existing industries and only a few can eventually create new market spaces, the conclusion is straightforward: both situations are empirically valid. Indirectly, this is exactly what this book suggested. Both Porter's competitive advantage and Kim and Maubourgne's value innovation approaches are not only right but also complementary. In the same way that the internal competences, based on the Resource-Based View (RBV) of the firm, complement the strategic positioning of Michal Porter, by presenting the rationales for the creation of new markets spaces, hence allowing some players to move beyond the realms of existing industries, BOS also complements Porter's positioning school. Indeed, the examples used throughout the book suggest the empirical relevance of Porter's and Kim and Mauborgne's approaches. External and internal constrains result in the scope for BOS to be more compact than the (traditional) search for competitive advantages in existing markets. For instance, when compared with some business operating in consumer markets, long value chains make more difficult for Tier 2 or 3 suppliers to value innovate and create new market spaces. Some, however, may be brave enough to venture into SVI strategies and create not only value for customers but also satisfy social and environmental demands.

**Strategy 5: Sustainable Value Innovation**

The fifth sustainability strategy extended the concept of value innovation to the realms of management of environmental and social issues in business. SVI laid the basis for the creation of new market spaces in alignment with the demands for corporate environmental and social responsibility. In order to present the logic embedded in SVI strategies, Chapter 7 went deep into the problems and potential solutions for the automobile industry, as well as the ultimate services it is supposed to deliver. The analysis of the scope for individual mobility identified marginal or untapped market spaces, which can be served by automakers or newcomers. The embedded logic in the analysis indicates the basic requirements of SVI strategies, depicted in Figure 8.2.

The point of departure for the creation of SVI is the identification of the ultimate service a product is supposed to deliver. In the case of individual-motorized mobility, it became patent that private ownership of cars is just one, among many possibilities. Empirical examples show that, when compared with private car ownership, it is possible to satisfy the demands for individual motoring more efficiently, at both

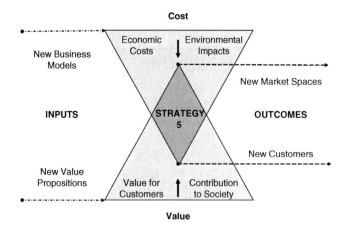

FIGURE 8.2 **Inputs and Outcomes of SVI**

lower costs and environmental impacts, as well as profitably in the short and long terms. For that to happen, however, innovative value propositions and/or business models are required from carmakers or any other suppliers of mobility services.

The solutions for sustainable mobility, currently in development by the car industry, focus on greener vehicles. Smaller cars, cleaner fuels and alternative powertrains are the main solutions pursued by the industry to reduce $CO_2$ emissions – the most pressing environmental demand of the industry nowadays. Although such developments should indeed compose the portfolio of solutions, they maintain intact most of the inefficiencies embedded in both the vehicle and its use. While the greening of motorization is essential to reduce total $CO_2$ emissions, according to the SVI logic systemic inefficiencies represent latent untapped market spaces. After all, if sustainable industries are supposed to emerge, efficiencies have to be optimized. As Chapter 7 explored in detail, this can be done not only profitably but also in a way that social inequalities and environmental impacts are reduced. More affordable services via Car-Sharing Organizations (CSOs), for instance, may allow current non-customers to become clients.

Overall, the deployment of SVI strategies depends less on technological breakthroughs than on management innovation.[3] Although technological innovation always helps, a bold strategy built mostly on managerial audacity and competence may be sufficient for the creation of SVI, as the case of Better Place indicates. After so many failures in introducing Electric Vehicles (EVs) to the marketplace, green entrepreneurship showed that, rather than technology, a major shift can be done in mobility patterns simply by adopting clever business models. Better Place suggests that companies can create value to existing and new customers at reduced economic and environmental costs. The case for car-sharing also suggested that, in order to align business needs with the gradual dematerialization of the economy, strategies based on SVI require competences for systemic innovation. In this respect, *Sustainability Strategies*, presents the main external (mostly systemic) factors that need to be addressed for the success of SVI.

## CONCLUSION

Translating sustainability into business strategies and practices remains a challenge for most companies. This is because sustainable development is more an inspiring vision than an articulated concept that can be put into practice. The Brundtland Report[4] presented the most commonly accepted definition of sustainable development: "the development that meets the needs of the present without compromising the ability of future generations to meet their own needs". The wide acceptance of this concept is due to the intrinsic win–win hypothesis embedded in it – that the satisfaction of current needs (winning today) can be achieved without compromising the satisfaction of future ones (winning tomorrow). Translating this concept into business terms can lead us to the following definition (among others) for sustainable business: the ability of firms to satisfy the economic needs of shareholders (private profits) without compromising nature and the needs of current and future generations (public benefits). In other words, under current economic norms, sustainable business is necessarily profitable business. This is the first aspect embedded in this book: that business profitability is the main concern of managers, even when dealing with social and environmental issues.

The second aspect relates to the role the book may have in promoting changes toward sustainability. Personally, most of us – including myself – would like the transformation to sustainable societies to happen faster than we are experiencing. Who would not endorse the possibility of creating wealth while preserving nature, today? We could all share the public benefits of having pure air, clean water and biodiversity preserved, now. Unfortunately, the complexity of human organization slows the pace of actions based on ecological principles. The road toward eco-economy[5] or natural capitalism[6] is much slower than most of us would like it to be. The very nature of open societies requires time for dialogue so that consensus can be reached for pacific social change. Although most of us are sympathetic to the move toward a society based on ecological values, such as proposed by deep ecologists,[7] this transition would still require the right incentives and some sort of management. Overall, the evolution toward more sustainable systems of production and consumption involves great degree of political negotiation, technological development and entrepreneurship. Even if we are able to deploy radically new technologies or business models, their implementation requires incremental institutional reform; or what we could call *radical reformism*.[8]

Together, these two aspects embedded in this book means it has been developed within the logic of commerce and open societies, in which wealth creation results from the negotiated expansion of both private profits and public benefits. Within this logic, I hope *Sustainability Strategies* contributes to the fastest possible pace of radical reformism in firms, industries and societies. By presenting a realistic view of what works or pays and what does not in the realms of corporate environmentalism, the book helps companies to be more effective and so to become increasingly ambitious with their sustainability strategies. Although I would rather prefer the scope for win–win scenarios to be broader than they are in practice, a realistic account is more useful than unfounded believes about reality. This is not to say that opportunities for the greening of industry do not exist. They do exist and they are vast. However, as the examples presented throughout the book suggested, opportunities are largely dispersed, confused by the context in which companies operate, and limited by the competences possessed by them. Within the existing rules, there is much companies can do for the creation of sustainable societies but, as this book suggests, we first need to have a better understanding of the conditions favoring or hindering companies to obtain returns from eco-investments. By doing so, we will eventually be able to identify the key elements making the capitalist enterprises truly sustainable.

# NOTES

## PREFACE

1   David Vogel, *The Market for Virtue: The Potential and Limits of Corporate Social Responsibility* (Washington, D.C.: Brookings Institution Press, 2006).

2   In this book, the term *sustainability* is used to designate the terms encompassed by the triple bottom line of economic, social and environmental sustainability. This is because academics, practitioners and the media have often used the terms environment, environmental, green and sustainability interchangeably. There is no assumption in this study that the term sustainability or sustainable development has been achieved by any contemporary organization, industry or society.

3   Michael Porter and Forest Reinhardt. "A Strategic Approach to Climate". *Harvard Business Review,* October, 2007, 22–26. See also: Michael Porter and Mark Kramer, "Strategy and Society: The Link Between Competitive Advantage and Corporate Social Responsibility," *Harvard Business Review,* December, 2006, 78–93; Michael Porter and Mark Kramer "The Competitive Advantage of Corporate Philanthropy," *Harvard Business Review,* 2002, 5–16.

4   Joan Magretta, *What Management Is: How It Works and Why It's Everyone's Business* (London: Profile Books, 2003).

5   Marc J. Epstein, *Making Sustainability Work: Best Practices in Managing and Measuring Corporate Social, Environmental and Economic Impacts* (Berrett-Koehler Publishers, 2008); Chris Laszlo, *Sustainable Value: How the World's Leading Companies Are Doing Well by Doing Good* (Sheffield: Greenleaf Publishing, 2008); Andrew J. Hoffman, *Competitive Environmental Strategy: A Guide to the Changing Business Landscape,* First Edition (Island Press, 2000).

6   For instance, see: William R. Blackburn, *The Sustainability Handbook: The Complete Management Guide to Achieving Social, Economic and Environmental Responsibility* (Earthscan Publications Ltd., 2007); Darcy Hitchcock and Marsha Willard, *The Business Guide to Sustainability: Practical Strategies and Tools for Organizations* (Earthscan Publications Ltd., 2006).

7   A specific approach for the implementation and control of corporate strategies, which has been extremely successful since the early 1990s, has been the Balanced Scorecard (BSC) approach, extensively explored by Robert Kaplan and David Norton. Their latest work include: Robert S. Kaplan and David P. Norton. *The Execution Premium: Linking Strategy to Operations for Competitive Advantage* (Boston, USA, Harvard Business School Press, 2008) and Robert S. Kaplan and David P. Norton, "Mastering the Management System," *Harvard Business Review,* 86, no. 1 (2008): 62–77; Robert Kaplan and David Norton, *Strategy Maps: Converting Intangible Assets Into Tangible Outcomes* (Boston, USA: Harvard Business

School Press, 2004). The BSC has also been adapted for the implementation of sustainability strategies. An early discussion about the elements involved in such adaptation was made by: Francesco Zingales, Anastasia O'Rourke and Renato J. Orsato, "Environment and Socio-related Balanced Scorecard: Exploration of Critical Issues," *INSEAD working papers*. 2002/47/CMER. See also: Frank Figge, Tobias Hann, Stefan Shaltegger, and Marcus Wagner, "The Sustainability Balanced Scorecard: Linking Sustainability Management to Business Strategy," *Business Strategy & Environment*, 11/5 (2002): 269–284; Andreas Moller and Stefan Schaltegger, "The Sustainability Balanced Scorecard as a Framework for Eco-Efficiency Analysis," *Journal of Industrial Ecology*, 9/4 (2005): 73–83. For the specific case of implementing Sustainable Value Innovation (SVI) Strategies, see: W. Chan Kim and Renée Mauborgne, *Blue Ocean Strategy: How To Create Uncontested Market space and Make Competition Irrelevant* (Boston, MA, USA: Harvard Business School Press, 2005).

## CHAPTER 1 – WHEN DOES IT PAY TO BE GREEN?

1 This was also known as the *Porter Hypothesis* in academic circles. The debate was triggered by the publication of the article: Michael Porter, "America's Green Strategy," *Scientific American*, 264 (1991): 96. The early stages of the debate can be found in: Richard A. Clarke *et al.*, "The Challenge of Going Green," *Harvard Business Review*, 72/4 (1994): 37–49; Noah Walley and Bradley Whitewead, "It's not Easy to Be Green," *Harvard Business Review*, 72/3: 46–52. See also Michael Porter & Clas Van der Linde, "Green and Competitive: Ending the Stalemate," *Harvard Business Review*, 73/5 (1995): 120–134; and the subsequent criticism of Karen Palmer, Wallace Oates, and Paul Portney, "Tightening Environmental Standards: The Benefits-Cost or the No-Cost Paradigm?" *Journal of Economic Perspectives*, 9/4 (1995): 119–132.

2 The term *environmental investment* (or eco-investment for short) relates to a wide array of practices that firms engage in to reduce direct or indirect environmental impacts of organizational processes, as well as impacts associated with the entire life cycle of products or services.

3 See, for instance: Andrew King and Michael Lenox, "Does It *Really* Pay to Be Green?" *Journal of Industrial Ecology*, 5/1 (2001): 105–116; Gerard J. Lewis and Neil Stewart, "The Measurement of Environmental Performance: An Application of Ashby's Law," *Systems Research and Behavioural Science*, 20 (2003): 31–52. Magnus Wagner, Stefan Schaltegger, and Walter Wehrmeyer, "The Relationship between the Environmental and Economic Performance of Firms: What Does Theory Propose and What Does Empirical Evidence Tell Us?" *Greener Management International*, 34 (2002): 95–108.

4 Forest Reinhardt, "Environmental Product Differentiation: Implications for Corporate Strategy," *California Management Review*, 40/4 (1998): 43–73.

5 Forest Reinhardt, "Market Failure and the Environmental Policies of Firms: Economic Rationales for 'Beyond Compliance' Behavior," *Journal of Industrial Ecology*, 3/1 (1999): 9–21.

6 Benjamin Bonifant, Matthew Arnold, and Frederick Long, "Gaining Competitive Advantage Through Environmental Investments," *Business Horizons*, 38/4 (1995): 37–48.

7 Michael Porter and Mark Kramer, "Challenging Assumptions," *European Business Forum*, Winter (2003): 3–4.

8 In this book I adopt the broad definition of *"ecological sustainability"* or *"ecological sustainable development"* proposed in the Brundtland Report (Report of the World

Commission on Environment and Development: "Our Common Future," 1987. (Also known as the Brundtland Report): "Sustainable development is the development that meets the needs of the present without compromising the ability of future generations to meet their own needs."

9   This section is based on a shortened version of the INSEAD Case 10/2007-5472 (2006), written by Renato J. Orsato and Fernando Von Zuben: Empowering the Bottom of the Pyramid via Product Stewardship: The Tetra Pak entrepreneurial networks in Brazil.

10  <www.unglobalcompact.org> September, 2008.

11  "NetAid: Educating, inspiring and empowering young people to fight global poverty." NetAid is an initiative of Mercy Corps. <www.netaid.org/> September, 2008.

12  <www.iblf.org> September, 2008.

13  <http://tetrapak.com.br/home.asp> September, 2008.

14  Tetra Pak Brazil invested around US$500,000/year during 1997–2000; US$1,000,000/year during 2001–2002; and US$2,000,000/year during 2003–2005 (source: Tetra Pak Brazil).

15  Klabin has been the focus of media and academic studies for its early commitments toward environmental protection. The company received several environmental awards, including "Prêmio Expressão de Ecologia", <www.expressao.com.br/ecologia/> for the years 1998, 1999, 2001, 2002 and 2005. See: <www.klabinonline.com.br/> September, 2008.

16  All values presented in this book have been converted from local currencies, such as the Brazilian Real and New Zealand and Australian Dollars, either to United States Dollars ($) and Euro (€), based on the average currency rates for September, 2008, according to: <www.x-rates.com> September, 2008. Since the numbers intend to give an idea of magnitude, they represent an approximate value of the original currency at the time of the publication of the sources.

17  Instituto de Pesquisas Technológicas de São Paulo.

18  Detailed information can be found in: Fernando von Zuben, "The Thermal Plasma Technology Separates Aluminum from Plastic in Packages," in *Proceedings of the International Conference on Energy, Environment and Disasters* (INCEED) (North Carolina, USA, 2005)

19  Megacities: São Paulo. *National Geographic.* 2005.

20  This section is based on a shortened version of the case written by Renato J. Orsato and Peter Wells: Eco-entrepreneurship: the bumpy ride of the Think. INSEAD Case 04/2008-5485.

21  The brand name of the product is *Th!nk* (with the exclamation mark instead of an 'i').

22  A *break-even point*, as a generic economic concept, represents the quantity where the contribution to fixed costs equals fixed costs. For the specific case of car manufacturing, the *break-even point* represents the amount of car production necessary to cover the fixed costs of production.

23  <http://think.no/> September, 2008.

24  The drawing of this figure was influenced by similar ones addressing corporate philanthropy, developed by Michael E. Porter and Mark R. Kramer, "The Competitive Advantage of Corporate Philanthropy," *Harvard Business Review*, 80/12 (2002): 56–69.

25  A good example of such approach is The Natural Step (TNS), developed by Karl-Henrik Robert, a Swedish physician. See, for instance: Brian Nattrass, *Dancing with the*

*Tiger: Learning Sustainability Step by Natural Step* (Gabriola Islands, Canada: New Society Publishers, 2002); Brian Nattrass, *The Natural Step for Business: Wealth, Ecology and the Evolutionary Corporation* (Philadelphia, Pa: New Society, 1999). See also; Peter M. Senge et al., *The Necessary Revolution: How Individuals and Organizations Are Working Together to Create a Sustainable World* (US Green Building Council, 2008); Maximilien Rouer and Anne Gouyon, *Réparer la planète: La révolution de l'économie Positive* (Jean-Claude Lattès, 2007)

26   Stewart Hart, "Beyond *Greening*: Strategies for a Sustainable World," *Harvard Business Review*, 75/1 (1997): 66–76.

27   Tom J. Brown and Peter Dacin, "The Company and the Product: Corporate Associations and Consumer Product Responses," *Journal of Marketing*, 61/1 (1997): 68–84.

28   For an overview of EPR, see: Thomas Lindhqvist, "Extended Producer Responsibility in Cleaner Production: Policy Principle to Promote Improvements of Product Systems" (PhD diss., IIIEE, Lund University, Sweden, 2000).

29   The term *base of the pyramid* has been coined by C. K. Prahalad and S. Hart, "The Fortune at the Bottom of the Pyramid," *Strategy + Business 26*, 2002 to refer to Tiers 4 and 5 of the population, with a buy power parity of $1500/year or below, which together count for 4 billion people.

30   The drawing of this figure was influenced by similar ones addressing corporate philanthropy, developed by Porter and Kramer, op. cit.

31   Renato J. Orsato, "Future Imperfect? The Enduring Struggle for Electric Vehicles," in *The Business of Sustainable Mobility*, ed. Paul Niewenhuis, Philip Vergragt and Peter Wells (London: Edward Elgar, 2006), 35–44.

32   An extensive review of the various approaches used in the *business and environment* literature can be found in: Luca Berchicci and Andrew King, "Chapter 11: Postcards from the Edge: A Review of the Business and Environment Literature," *The Academy of Management Annals*, 1/1 (2007): 513–547.

33   Normally out of the EPA's Toxic Release Inventory.

34   Andrew King and Michael Lenox, "Industry Self-Regulation without Sanctions: The Chemical Industry's Responsible Care Program," *The Academy of Management Journal*, 43/4 (2000): 698–716.

35   Torbjörn Brorson, *Environmental Management: How to Implement an Environmental Management System Within a Company or Other Organisation*, 3 éd. (Stockholm: EMS, 1999).

36   Kathleen M. Eisenhardt, "Building Theories from Case Study Research," *Academy of Management Review*, 14/4 (1989): 532–550.

37   Robert Stake "Case Studies," in *Strategies of Inquiry*, ed. Norman K. Denzin and Yvonna S. Lincoln (Thousand Oaks, CA: Sage, 1994).

38   C. Eden and C. Huxham, "Action Research for the Study of Organizations," in *Handbook of Organizational Studies*, ed. S. Clegg, C. Hardy, and W. Nord (London: Sage,1996), 565–580.

39   Data was drawn from empirical research developed during 2004–2007, as part of the project called "*Strategic Environmental Management in European and Autralasian Firms*," awarded by the European Commission, Marie Curie Actions (MOIF-CT-2004-509911). Selected cases have also been chosen from data collected during the period 1999–2004, as part of the action-research program with 35 Swedish companies, at the

International Institute for Industrial Environmental Economics (IIIEE), as well as the research about the global automobile industry, presented in: Renato J. Orsato, "The Ecological Modernization of Industry: Developing Multi-disciplinary Research on Organization & Environment" (PhD diss., University of Technology, Sydney, Australia: 2001).

40  W. Chan Kim and Renée Mauborgne, *Blue Ocean Strategy: How to Create Uncontested Market Space and Make Competition Irrelevant* (Boston, MA, USA: Harvard Business School Press, 2005).

## CHAPTER 2 – WHAT ARE SUSTAINABILITY STRATEGIES?

1  A generic overview of the various schools of strategic management is presented by Henry Mintzberg, Bruce Ahlstrand, and Joseph Lampel, *Strategy Safari: A Guided Tour Through the Wilds of Strategic Management* (New York: The Free Press, 1998).

2  For instance: *Business Strategy & the Environment Journal, Corporate Environmental Strategy Journal, Journal of Industrial Ecology.*

3  Forest L. Reinhardt, *Down To Earth: Applying Business Principles to Environmental Management* (Boston, USA: Harvard Business School Press, 2000).

4  See chapter 4 in Daniel Esty and Andrew Winston, *Green to Gold: How Smart Companies Use Environmental Strategy to Innovate, Create Value, and Build Competitive Advantage* (New Haven, USA: Yale University Press, 2006); see also: chapter 4 in Reinhardt, op. cit.

5  Michael Porter, "What Is Strategy?" *Harvard Business Review*, 74/6 (1996): 61–79, 61–62.

6  Joan Magretta, *What Management Is: How It Works and Why it's Everyone's Business* (London: Profile Books, 2003).

7  W. Chan Kim and Renée Maubourgne, "Value Innovation: The Strategic Logic of High Growth," *Harvard Business Review*, 82/7–8 (2004): 172–180.

8  W. Chan Kim and Renée Mauborgne, *Blue Ocean Strategy: How to Create Uncontested Market Space and Make the Competition Irrelevant* (Boston, MA: Harvard Business School Press, 2005).

9  Michael Porter, *Competitive Strategy: Techniques for Analyzing Industries and Competitors* (New York: The Free Press, 1980).

10  Michael Porter, "What Is Strategy?" *Harvard Business Review*, 74/6 (1996): 61–79, 61–62.

11  For an overview of the sources of competitive advantage, see: chapter 2 in Michael Porter, *Competitive Advantage: Creating and Sustaining Superior Performance* (London: Free Press, 1985).

12  Michael Porter, "Towards a Dynamic Theory of Strategy," *Strategic Management Journal* (1991): 95–118.

13  Sandra A. Waddock, Charles Bodwell, and Samuel B. Graves, "Responsibility: The New business Imperative," *Academy of Management Executive*, 16/2 (2002): 132–148; Sandra A. Waddock and Charles Bodwell, *Total Responsibility Management: The Manual* (Sheffield: Greenleaf Publishing, 2007).

14 Michael Berry and Denis Rondinelli, "Proactive Corporate Environmental Management: A New Industrial Revolution," *Academy of Management Executive*, 12/2 (1998): 38–50; Forest Reinhardt, "Market Failure and the Environmental Policies of Firms: Economic Rationales for *Beyond Compliance* Behavior," *Journal of Industrial Ecology*, 3/1 (1999): 9–21.

15 Sandra A. Waddock and Charles Bodwell, "Managing Responsibility: What Can Be Learned from the Quality Movement?" *California Management Review*, 47/1 (2004): 25–37.

16 Work on the *Resource-Based View of the Firm* in strategic management began with the articles of: (i) Birger Wernefelt, "A Resource-Based View of the Firm," *Strategic Management Journal*, 5 (1984): 171–180; (ii) Richard Rumelt, "Toward a Strategic Theory of the Firm," in *Competitive Strategic Management*, ed. Englewood Cliffs (NJ: Prenctice Hall, 1984); and (iii) Jay Barney, "Strategic Factor Markets: Expectations, Luck, and Business Strategy," *Management Science*, 32/10 (1986): 1231–1242. A *situational* review of the Resource-Based View can be found in: Jay Barney and William Hesterly, "Organizational Economics: Understanding the Relationship Between Organizations and Economic Analysis," in *Handbook of Organization Studies*, ed. Stewart Clegg, Cynthia Hardy, and Walter Nord (London: Sage, 1996): 115–147.

17 The concept of *organizational processes* employed here encompasses both the activities of controlling manufacturing, *production*, or *industrial processes*, as well as generic *management processes*, which relate mainly but are not limited to (bureaucratic) activities performed by the various members of an organization. For an overview of classic organizational processes, see: Richard Hall, *Organizations: Structures, Processes, and Outcomes* (Upper Saddler River: Prentice Hall, 1999).

18 This section is based on Renato J. Orsato, "Competitive Environmental Strategies: When Does It Pay to Be Green?" *California Management Review*, 48/2 (2006): 127–141.

19 For instance: Pieter Winsemius and Ulrich Guntran, *A Thousand Shades of Green: Sustainable Strategies for Competitive Advantage* (London: Earthscan, 2002); Bob Williard, *The Sustainability Advantage: Seven Business Case Benefits of a Triple Bottom Line* (Gabriola Island, Columbia, Canada: New Society Publishers, 2002); Paul Hawken, Amory Lovins, and Hunter Lovins, *Natural Capitalism: The Next Industrial Revolution* (London: Earthscan, 1999).

20 Orsato op. cit. Figure p. 131.

21 For an overview of *Voluntary Agreements* and how they may influence firm's competitiveness, see: Magali Delmas and Ann K. Terlaak, "A Framework for Analyzing Environmental Voluntary Agreements," *California Management Review*, 43/3 (2001): 44–61.

22 Such as those presented by Stewart Hart, "Beyond Greening: Strategies for a Sustainable World," *Harvard Business Review*, 75/1 (1997): 66–76; and Dexter Dunphy and Andrew Griffiths, *Sustainable Corporation: Organizational Renewal in Australia* (Frenchs Forest, Australia: Allen & Urwin, 1998), among others.

23 Christopher B. Hunt and Ellen R. Auster, "Proactive Environmental Management: Avoiding the Toxic Trap," *Sloan Management Review*, 31/2 (1990): 7–19.

24 Stuart L. Hart, *Capitalism at the Crossroads: Aligning Business, Earth, and Humanity*, 2nd edn. (Wharton, Pennsylvania: Wharton School Publishing, 2007).

25 Porter, op. cit.

26  In 2006 GE was the fourth most valuable corporate brand in the world, worth $48,907 Million Interbrand Corporation | Global Branding Consultancy. <www.interbrand.com/> September, 2008.

27  Charles Corbett and David Kirsh, "ISO 14000: an agnostic's report from the front line," *ISO 9000 + ISO 14000 News* 2 (2000): 4–17. Richard Florida and Derek Davidson, "Gaining from Green Management: Environmental Management Systems Inside and Outside the Factory," *California Management Review*, 43/3 (2001): 64–85.

28  Kim and Mauborgne, op. cit.

29  This figure is based on the logic of *value innovation* presented in Page 16 of Kim and Mauborgne, op. cit.

## CHAPTER 3 – ECO-EFFICIENCY

1  Michael Porter, *The Competitive Advantage of Nations* (London: The Macmillan Press, 1990).

2  Michael Porter and Clas Van der Linde, "Green and Competitive: Ending the Stalemate," *Harvard Business Review*, 73 (1995): 120–134.

3  Paul Hawken, Amory Lovins, and Hunter Lovins, *Natural Capitalism: The Next Industrial Revolution* (London: Earthscan, 1999). See also previous work on resource productivity developed by E. von Weizsäcker, Amory Lovins and Hunter Lovins, *Factor Four: Doubling Wealth – Halving Resource Use* (Sydney: Allen & Unwin, 1997).

4  For a review of the concept of eco-efficiency, see: Chris Ryan, *Digital Eco-Sense: Sustainability and ICT – A New Terrain for Innovation* (Melbourne, Australia: Lab 3000, 2004).

5  Charles J. Corbett and Robert D. Klassen, "Extending the Horizons: Environmental Excellence as Key to Improving Operations," *Manufacturing & Service Operations Management*, 8/1 (2006): 5–22; present rationales for companies to consider the environmental excellence of entire supply chain as means of reaching improvements in the operations of firms.

6  James P. Womack and Daniel T. Jones, *Lean Thinking: Banish Waste and Create Wealth in Your Corporation* (London: Simon & Schuster, 2003).

7  ZERI is a network organization that researches and fosters industrial eco-efficiency, which emerged out of a special program of the United Nations University, Tokyo, Japan.

8  The brewery's outputs can also include a bio-digester to treat the sewage and transform it into composted fertilizer and methane. The methane can replace a portion of the fuel needed for the brewery's boiler. Algae basins can also be used to process water from the digester and grow algae for cattle feed. Finally, the water from the basins can flow into deepwater ponds to grow fish. <www.zeri.org/> October, 2008.

9  Gunter Pauli, *Upsizing: The Road to Zero Emissions – More Jobs More Income and No Pollution* (Sheffield: Greenleaf, 1998).

10  C. K. Prahalad and Gary Hamel, "The Core Competence of the Corporation," *Harvard Business Review*, 68/3 (1990): 79–92.

11  According to Michael Pollan, feeding large quantities of grains to cows might make business sense but it can be detrimental to their health. Cows are ruminants which eat mainly grass. Although small amounts of cereals can be added to their diet, large quantities

cause digestive problems. See: Michael Pollan, *The Omnivore's Dilemma. The Search for the Perfect Meal in a Fast-Food World* (London: Bloomsbury Publishing, 2006).

12  Heineken International Group Corporate Relations, personal communication on July 11, 2007.

13  Qinghua Zhu *et al.* "Industrial Symbiosis in China: A Case Study of the Guitang Group," *Journal of Industrial Ecology*, 11/1 (2007): 31–42. See also; Qinghua Zhu and Raymond P Cote, "Integrating Green Supply Chain Management into an Embryonic Eco-Industrial Development: A Case Study of the Guitang Group," *Journal of Cleaner Production*, 12/8–10 (2004): 1025–1035.

14  An early review of the main concepts and research areas of industrial ecology can be found in Frank den Hond, "Industrial Ecology: A Review," *Regional Environmental Change*, 1/2 (2000): 60–69; More recent developments can be found in the *Journal of Industrial Ecology*. <www.blackwellpublishing.com/journal.asp?ref=1088-1980> October, 2008.

15  An analysis of potential for industrial ecosystems can be found in: Catherine Hardy and Thomas E. Graedel, "Industrial Ecosystems as Food Webs," *Journal of Industrial Ecology*, 6/1 (2002): 29–38. See also; Robert U. Ayres, "Creating Industrial Ecosystems: A Viable Management Strategy?" *International Journal of Technology Management*, 12/5,6 (1996): 608.

16  A thorough assessment of the Kalundborg eco-park is presented by: Noel B. Jacobsen, "Industrial Symbiosis in Kalundborg, Denmark: A Quantitative Assessment of Economic and Environmental Aspects," *Journal of Industrial Ecology*, 10/1–2 (2006): 239–255.

17  A broad assessment of eco-industrial parks has been developed by: D. Gibbs, P. Deutz, and A. Procter, "Industrial Ecology and Eco-Industrial Development: A New Paradigm for Local Regional Development?" *Regional Studies*, 39/2(2005): 171–183. A more recent assessment can be found in: Marian R. Chertow, "Uncovering Industrial Symbiosis," *Journal of Industrial Ecology*, 11/1 (2007): 11–30; and M. Mirata, "Industrial Symbiosis – A Tool for More Sustainable Regions" (PhD diss., IIIEE, University of Lund, 2005).

18  An overview of similarities and difference between natural and industrial systems can be found in D. Richards and G. Pearson, *The Ecology of Industry: Sectors and Linkages* (Washington DC: National Academy Press, 1998).

19  Marian R. Chertow, "Uncovering Industrial Symbiosis," *Journal of Industrial Ecology*, 11/1 (2007): 11–30.

20  Details of the synergies in Kwinana can be found in: Dick van Beers, Status Report on Regional Synergies in the Kwinana Industrial Area (Perth, WA, Australia: Centre for Sustainable Resource Processing, 2006); See also: Dick van Beers, Glen Corder, Albena Bossilov, and Rene van Berkel, "Industrial Symbiosis in the Australian Mineral Industry: The Case of Kwinana and Gladstone," *Journal of Industrial Ecology*, 11/1 (2007): 55–72.

21  An excellent account of EIPs planned and under development can be found in: "Welcome to Indigo Development," <www.indigodev.com/index.html> September, 2008. See also: Nelson Nemerw, *Zero Pollution for Industry: Waste Minimization Through Industrial Complexes* (New York: Wiley, 1995).

22  Jose, Goldemberg "The Ethanol Program in Brazil," *Environmental Research Letters*, Institute of Physics Publishing edition (2006): sec. Lett: 1.

23  Petri Ristola and Murat Mirata, "Industrial Symbiosis for More Sustainable, Localised Industrial Systems," *Progress in Industrial Ecology – An International Journal*, 4/3,4 (2007): 184–202.

24 There is a plethora of books, articles and case studies explaining the process that culminated in the Kyoto Protocol and its mechanisms. See, for instance: Robert Henson, *The Rough Guide to Climate Change: The Symptoms, the Science and the Solutions* (New York: Rough Guides Ltd., 2006).

25 Carbon dioxide ($CO_2$), methane ($CH_4$), nitrous oxide ($N_2O$), sulfur hexafluoride ($SF_6$), hydrofluorcarbons ($HFC_s$) and perfluorcarbons ($PFC_s$).

26 <http://unfccc.int/essential_background/convention/items/2627.php/> October, 2008.

27 There are four IPCC assessment reports: the 1st published in 1990, the 2nd in 1995, the 3rd in 2001 and the 4th in 2007.

28 For detailed information see "UNFCCC" at: <http://unfccc.int/kyoto_protocol/mechanisms/items/1673.php/> October, 2008.

29 Besides EU allowances (EUAs) allocated in National Allocation Plans (NAP), the commodities generated by the Clean Development Mechanisms (Certified Emission Reductions – CERs) and Joint Implementation (Emission Reduction Unit – ERUs) are allowed for compliance use in EU ETS. However, the amount of certificates from project-based instruments is limited by the NAP. See Directive 2004/101/EC about "linking" EU ETS with Kyoto project-based mechanisms.

30 €9.4 billion were traded in 2005 and €22.5 billion in 2006, but because the European Commission was too generous in Phase 1 allowances (2005–2008), by July of 2007 the price of carbon credits was virtually worthless. Having learnt the lesson, the EC was more restrictive with its allowances in phase 2 (2008–2012). By October, 2008, carbon credits were worth €23/tone of $CO_2$-equivalent. <www.pointcarbon.com/> October, 2008.

31 The basic commodity of the Kyoto protocol for trading is called "assigned amount units" (AAUs). Clean Development Mechanism uses Certified Emission Reductions (CER), and Joint Implementation uses Emission Reduction Unit (ERU). Although there are different names for the commodities, they all represent 1 ton of $CO_2$ reduction.

32 More precisely, in 2008 there were 3,967 CDM projects. <www.cdmpipeline.org/cdm-projects-type.htm> October, 2008.

33 In 2006 the CDM market was already worth €5 billion, with more than 200 projects approved in India, China and Brazil. By July, 2007, China took the lead and alone accounted for 61 percent of the CDM market, with 524 projects approved. In 2008, China represented the 52 percent of the expected average annual CERs, India represented 13 percent followed by Brazil at around 9 percent. <http://cdm.unfccc.int/Statistics/index.html> October, 2008.

34 Cleaning Up: A Special Report on Business and Climate Change, *The Economist*, June 2nd, 2007.

35 Veja. "Salvar o Planeta Da Lucro," December 6, 2006, pages 116–119 Magazine.

36 <www.sapaulista.com.br/> October, 2008.

37 Andrew Hoffman, "Climate Change Strategy: The Business Logic Behind Voluntary Greenhouse Gas Reductions," *California Management Review*, 47/3 (2005): 21–46.

38 Ans Kolk and Jonatan Pinkse, "Business Responses to Climate Change: Identifying Emergent Strategies," *California Management Review*, 47/3 (2005): 6–20.

39 This is clearly reflected in the title of the book: Andrew J. Hoffman and John G. Woody, *Climate Change: What's Your Business Strategy?* (Harvard Business School Press, 2008).

40  Daniel C. Esty and Michael E. Porter, "Industrial Ecology and Competitiveness: Strategic Implication for the Firm," *Journal of Industrial Ecology*, 2/1 (1998): 35–43.

41  Hawken, Lovins, and Lovins, op. cit.

42  Kolk and Pinkse op. cit.

43  An account of this transformation can be found in Pollan op. cit.

44  L. Nilesen and T. Jeppesen "Tradable Green Certificates in Selected European Countries – Overview and Assessment", *Energy Policy*, 31/I, (2003): 3–14," *Fuel and Energy Abstracts*, 44/4 (July, 2003): 270.

45  An in depth analysis of Tradable White Certificates is presented by: L. Mundaca, "Markets for Energy Efficiency: Exploring the New Horizons of Tradable Certificate Schemes" (PhD diss., IIIEE, Lund University, 2008).

46  Mundaca op. cit.

47  Mundaca op. cit.

48  Hawken, Lovins, and Lovins, op. cit.

49  Emissions Trading: Lightly Carbonated. *The Economist*, August 4th, 53 (2007).

50  Luis Mundaca, "Transaction Costs of Tradable White Certificate Schemes: The Energy Efficiency Commitment as Case Study," *Energy Policy*, 35/8 (2007): 4340–4354.

## CHAPTER 4 – BEYOND COMPLIANCE LEADERSHIP

1  For a complete description of the case, see: Renato J. Orsato and Kes McCormick: *Eco-activism Greenpeace, the Oil Industry and the Stuart Oil Shale Project in Australia*. INSEAD case: 11/2006–5339.

2  This information is based on a document that was leaked to Greenpeace by Environment Australia and Australian Greenhouse Office (2002). Information about this can be found at: <http://www.greenpeace.org/australia/news-and-events/news/Climate-change/shale-oil-plant-shut-down-afte> October, 2008.

3  Oil shale becomes competitive with oil prices above 70US$ per barrel. See Colin J. Campbell and Jean H. Laherrère, "The End of Cheap Oil," *Scientific American*, 278/3 (1998): 60–65.

4  Crimes of Protest, *Time Magazine*, August 28 (1995).

5  A basic explanation of the role of *agency* in business-environment relationships is provided by: Renato J. Orsato and Stewart R. Clegg, "The Political Ecology of Organizations: Toward a Framework for Analyzing Business-Environment Relationships," *Organization & Environment*, 12/3 (1999): 263–279.

6  In 1995 *Shell* wanted to sink obsolete oil rigs in the North Sea but the subsequent outcry lead by *Greenpeace* enticed consumers to boycott Shell petrol, resulting in a 60 percent downfall in sales in Germany alone. The pressure from consumers and the general public induced Shell to shift its strategy and the oil platform, Brent Spar, instead of being sunk in deep-sea waters, was dismantled on land. See: Lisa Dickson and Alistar McCulloch, "Shell, The Brent Spar and *Greenpeace*: A Doomed Tryst?" *Environmental Politics*, 5 (1996). See also: David Vogel, *The Market for Virtue: The Potential And Limits of Corporate Social Responsibility* (Washington, D.C.: Brookings Institution Press, 2006).

7   C. N Smith, "Consumers as Drivers of Corporate Social Responsibility," in *The Oxford Handbook of Corporate Social Responsibility*, ed. Andrew Crane, Abagail McWilliams, Dirk Matten, Jeremy Moon, and Donald Siegel (Oxford: Oxford University Press, 2008): 281–302.

8   Mark Haugaard, *The Constitution of Power: A Theoretical Analysis of Power, Knowledge and Structure* (Manchester: Manchester University Press, 1997).

9   C. Hillenbrand and K. Money "Corporate Responsibility and Corporate Reputation," *Corporate Reputation Review*, 10/4 (2007): 261–277.

10  Hillenbrand and Money op.cit.

11  The concept of Green Clubs has been proposed by: Aseen Prakash and Matthew Potoski, *The Voluntary Environmentalists: Green Clubs, ISO 14001, and Voluntary Environmental Regulations* (Cambridge: Cambridge University Press, 2006). Many Green Clubs also include corporate social responsibilities and broader human rights issues, such as presented by the UN Global Compact.

12  Andrew King and Michael Lenox, "The Strategic Use of Decentralized Institutions: Exploring Certification with the ISO 14001 Management Standard," *Academy of Management Journal*, 48/6 (2005): 1091–1106. See also: P. Ingram and B. Silverman "The New Institutionalism in Strategic Management" in *Advances in Strategic Management*, ed. P. Ingram & B. S. Silverman (Greenwich: CT: JAI Press): 1–32.

13  Rupert Howes, Jim Skea, and Bob Whelan, *Clean and Competitive? Motivating Environmental Performance in Industry* (London: Earthscan, 1997).

14  Jennifer Nash and John Ehrenfeld, "Codes of Environmental Practice: Assessing Their Potential as a Tool for Change," *Annual Review of Energy and the Environment*, 22 (1997): 487–535.

15  Prakash and Potoski, op. cit.: 47.

16  A useful review of the main Green Clubs and a pletora of other VEIs is presented by: Sandra Waddock, "Building a New Institutional Infrastructure for Corporate Responsibility," *Academy of Management Perspectives*, 22/3 (2008): 87–108.

17  Approximately 2500 people died, and 200,000 were injured as a consequence of the accident on December 3, 1984. For details see: Paul Shrivastava, "Corporate Self-*Green*wal: Strategic Responses to Environmentalism," *Business Strategy and the Environment*, 1/3, (1992): 9–21; Paul Shrivastava, "Ecocentric Management for a Risk Society," *Academy of Management Review*, 20/1 (1995): 118–137.

18  Nash and Ehrenfeld op. cit.; Jennifer Howard, Jennifer Nash, and John Ehrenfeld, "Industry Codes as Agents of Change: Responsible Care Adoption by U.S. Chemical Companies," *Business Strategy and the Environment*, 8/5 (1999): 281–295; King and Lenox op. cit.

19  <www.icmm.com/> September, 2008.

20  <www.ceres.org/> September, 2008.

21  A debate on Civil Regulation can be found in Simon Zadek, *The Civil Corporation: The New Economy of Corporate Citizenship* (London: Earthscan, 2006). See also: David Vogel op. cit., and chapter 11 of Andrew Crane and Dirk Matten, *Business Ethics: Managing Corporate Citizenship and Sustainability in the Age of Globalization*, 2nd ed. (Oxford: Oxford University Press, 2007).

22 Coca-Cola joined CERES in 1998. Later it also engaged in a partnership with WWF to conserve the world's sweet-water reserves. The Company is also listed in the FTSE Good and the Dow Jones Sustainability indexes. It has also developed an environmental management system called eKo. <www.coca-cola.com/> September, 2008.

23 <http://www.globalreporting.org/Home> September, 2008.

24 The WBCSD was formed through a merger between the Business Council for Sustainable Development, in Geneva, and the World Industry Council for the Environment, an initiative of the International Chamber of Commerce (ICC), in Paris. See: <www.wbcsd.ch/templates/TemplateWBCSD5/layout.asp?MenuID=1> September, 2008.

25 <www.unglobalcompact.org/> September, 2008.

26 UN Global Compact Annual Progress survey 2007 produced in cooperation with Wharton, pages 6–25.

27 UN Global Compact Annual Progress survey 2007 produced in cooperation with Wharton, page 14.

28 Charles J. Corbett and Michael V. Russo, *Special Report – The Impact of ISO 14001: ISO 14001: Irrelevant or Invaluable?*, ISO Management Systems.

29 Council Regulation (EEC) No 1836/93 of 29 June, 1993 allowing voluntary participation by companies in the industrial sector in a Community eco-management and audit scheme.

30 The TQM "movement" has been initiated by W. Edwards Deming and Joseph Duran (both Americans), who spread the ideas of *defect prevention* through the control of production and businesses processes and the elimination of causes of major variations of quality in products or services commercialized by firms. Because of its practical and academic ramifications in the 1980s and first half of 1990s, TQM became one of the most influential management phenomena of the second half of the 20th century. During that period, a large number of handbooks and academic papers were published on the topic. Today, TQM principles and methods are part of the normal curriculum in management schools and textbooks. For an overview of quality and quality principles in management, see: A. Blanton Godfrey and Joseph M. Juran, *Juran's Quality Handbook* (USA: McGraw-Hill, 1998).

31 Robert Kaplan and David Norton, *Strategy Maps: Converting Intangible Assets into Tangible Outcomes* (Boston, USA: Harvard Business School Press, 2004); Research in this area has tried to link environment-related intangibles to the financial bottom line: See, for instance: Frank Figge, Tobias Hann, Stefan Shaltegger, and Marcus Wagner, "The Sustainability Balanced Scorecard: Linking Sustainability Management to Business Strategy," *Business Strategy & Environment*, 11/5 (2002): 269–284. See also: Francesco Zingales, Anastasia O'Rourke, and Renato J. Orsato, "Environment and Socio-Related Scorecard: Exploration of Critical Issues," *INSEAD Working Paper* (2002/47/CMER).

32 Joan Magretta, *What Management Is: How It Works and Why It's Everyone's Business* (London: Profile Books, 2003) offers an excellent account of *value creation* in organizations.

33 Chad Nehrt, "Maintainability of First Mover Advantages When Environmental Regulations Differ Between Countries," *Academy of Management Review*, 23/1 (1998): 77–97.

34 Interview with Mr Richard Riddiford, Managing Director of Pallister on 22/04/2005. See also: Michael Beverland, Brand Value, Convictions, Flexibility, and New Zealand Wine, *Business Horizons*, 47/5 (2004): 53–61.

35  In 2008, the *Boticário* had more than 900 franchisees in Brazil. The company won several national and international ecological prizes.

36  The *Boticário Nature Protection Foundation* sponsors around US$500,000 yearly in conservation projects. In February, 1995, the foundation inaugurated a 17,000 hectares park of native forest in the State of Paraná, Brazil.

37  The information presented here was obtained through interviews with the Vice-President of the *Boticário* and the director of the *Boticário Nature Protection Foundation*.

38  <www.pewclimate.org/companies_leading_the_way_belc> April, 2008.

39  The Climate savers Programme General Fact sheet downloaded from <www.worldwildlife.org> April, 2008.

40  <www.worldwildlife.org/climate> April, 2008.

41  The Climate savers Programme General Fact sheet downloaded from <www.worldwildlife.org> April, 2008.

42  <www.edf.org/page.cfm?tagID=82> April, 2008.

43  <www.epa.gov/stateply/> April, 2008.

44  <www.theclimategroup.org/index.php/about_us/> April, 2008.

45  Alcan Inc., Alcoa, American International Group, Inc. (AIG), Boston Scientific Corporation, BP America Inc., Caterpillar Inc., Chrysler LLC, ConocoPhillips, Deere & Company, The Dow Chemical Company, Duke Energy, DuPont Environmental Defense, Exelon Corporation, Ford Motor Company, FPL Group, Inc., General Electric, General Motors Corp. Johnson & Johnson, Marsh, Inc., National Wildlife Federation, Natural Resources Defense Council, The Nature Conservancy, NRG Energy, Inc., PepsiCo, Pew Center on Global Climate Change, PG&E Corporation, PNM Resources, Rio Tinto, Shell, Siemens Corporation, World Resources Institute, Xerox Corporation.

46  Andrew J. Hoffman, "If You're Not at the Table, You're on the Menu," *Harvard Business Review,* 85/10 (2007): 34–35.

47  Michael Porter & Clas Van der Linde, "Green and Competitive: Ending the Stalemate," *Harvard Business Review,* 73/5 (1995): 120–134.

48  Andrew J. Hoffman, "Climate Change Strategy: The Business Logic behind Voluntary Greenhouse Gas Reductions," *California Management Review,* 47/3 (2005): 21–46.

49  HM Treasury, Stern Review on the Economics of Climate Change, 2007; IPCC, IPCC Fourth Assessment Report "Mitigation of Climate Change", 2007, <http://www.ipcc.ch/ipccreports/ar4-wg3.htm>, September, 2008.

50  Making things worse, Bjorn Lomborg, a Danish populist, has been doing a disservice to business and society by using the complexities involved in the climate issue to further confuse the public and create media space for himself. By proposing intra- and inter-generational trade-offs between social and environmental issues, Lomborg create ate space for self-promotion (private profits) in detriment of any public benefit. Bjorn Lomborg, *Cool It: The Skeptical Environmentalist's Guide to Global Warming* (Vintage, 2008).

51  Prakash and Potoski, op. cit.

52  Forest Reinhardt, *Down to Earth: Applying Business Principles to Environmental Management* (Boston, USA: Harvard Business School Press, 2000).

53 Petra Christmann and Glen Taylor, "Globalization and the Environment: Strategies for International Voluntary Environmental Initiatives," *Academy of Management Executive,* 16/3 2002: 121–135.

54 Glen Dowel, Stuart Hart and Bernard Yeung, "Do Corporate Global Environmental Standards Create or Destroy Value?," *Management Science,* 46/8 (2000): 1059–1074.

55 Jill Meredith Ginsberg and Paul Bloom, "Choosing the Right Green Marketing Strategy," *MIT Sloan Management Review,* 46/1 (2004): 79–84.

56 In the years 2000 to 2009, the value of the Coca-Cola brand has situated between $65–70 billion. See: <www.interbrand.com> September, 2008.

57 Marc Gunther, "Coca-Cola's Green Crusader – It helps when the CEO is committed to sustainability", *Fortune Magazine,* April 28 (2008): 68.

58 Quite often natural gas is portrayed as a cleaner fuel, but this is only because it is compared with more polluting types of hydrocarbons, such as petrol or diesel.

59 Corbett and Russo op. cit.

60 Andrew King, Michael Lenox, and Ann Terlaak, "The Strategic Use of Decentralized Institutions: Exploring Certification with the ISO 14001 Management Standard," *Academy of Management Journal,* 48/6 (2005): 1091–1106.

61 Chad Nehrt, "Maintainability of First Mover Advantages When Environmental Regulations Differ Between Countries," *Academy of Management Review,* 23/1 (1998): 77–97.

62 Studies about the diffusion of environmental standards can be found in: Paulo Albuquerque, Bart J. Bronneberg, and Charles Corbett, "A Spaciotemporal Analysis of the Global Diffusion of ISO 9000 and 14000 Certification," *Management Science,* 53/3 (2007): 451–468; Magali Delmas and Ivan Montiel, "The Diffusion of Voluntary International Management Standards: Responsible Care, ISO 9000, and ISO 14001 in the Chemical industry," *Policy Studies Journal,* 36/1 (2008): 65–93.

63 Petra Christmann and Glen Taylor, "Globalization and the Environment: Determinants of Firm Self-regulation in China," *Journal of International Business Studies,* 32/3 (2001): 439–458.

64 Paulo Albuquerque, Bart J. Bronneberg, and Charles Corbett (2007) op. cit.

65 Ginsberg and Bloom, op. cit.

66 BP Sustainability Report, 2005, 26.

67 The Index was created 1999 and was followed by a European version FTSE4GoodEurope 50 in October, 2001. The Index rates companies by assessing their environmental management systems (EMS), eco-design efforts, corporate governance mechanisms and employment policies, among other aspects. See: <www.sustainability-index.com/> April, 2008.

68 Matten and Crane, op. cit.: 214–215.

## CHAPTER 5 – ECO-BRANDING

1 Michael Porter, *Competitive Advantage: Creating and Sustaining Superior Performance* (London: Free Press, 1985): 120.

2  This section is based on James Austine and Ezequiel Reficco "Forest Stewardship Council," Harvard Business School Case 9–303–047 (Boston, MA: Harvard Business School Publishing, 2006); David Vogel, *The Market for Virtue: The Potential and Limits of Corporate Social Responsibility* (Washington, D.C: Brookings Institution Press, 2006): 114–121; and information collected in the web-site of Forest Stewardship Council. <www.fsc.org> June, 2008.

3  Austine and Reficco, op. cit., Appendix 1.

4  For an overview of the initial years of the Sustainable Forestry Initiative, see Forest L Reinhardt, *Down to Earth: Applying Business Principles to Environmental Management* (Boston, MA: Harvard Business School, 1999): 55–57.

5  Petra Christmann and Glen Taylor, "Globalization and the Environment: Strategies for International Voluntary Environmental Initiatives," *Academy of Management Executive*, 16/3(2002): 121–135.

6  Christmann and Taylor op.cit.

7  Austine and Reficco, op. cit., page 14.

8  FSC <www.fsc.org> May, 2008.

9  Vogel op. cit.: 118; Joseph Domask, "From Boycotts to Global Partnership; NGO's, the Private Sector and the Struggle to protect the World's Forest," in *Globalization and NGOs: Transforming Business, Government, and Society*, eds. Jonathan Doh and Hildy Teegen (Westport, CN.: Praeger, 2003): 168.

10  The International Organization for Standardization (ISO) has the following standards for environmental labels and declarations: ISO 14020:2000 (General principles); ISO 14021:1999 (Type II environmental labelling); ISO 14024:1999 (Type I environmental labelling); ISO 14025:2006 (Type III environmental declarations). <www.iso.org> May, 2008.

11  According to the Roper survey [Roper ASW, "Green Gauge Report 2002" (New York: Roper ASW, 2002)], there are five types of consumers: "*True – Blue Greens* (9%): True Blues have strong environmental values and take it upon themselves to try to effect positive change. They are over four times more likely to avoid products made by companies that are not environmentally conscious. *Greenback Greens* (6%): Greenbacks differ from True Blues in that they do not take the time to be politically active. But they are more willing than the average consumer to purchase environmentally friendly products. *Sprouts* (31%): Sprouts believe in environmental causes in theory but not in practice. Sprouts will rarely buy a green product if it means spending more, but they are capable of going either way and can be persuaded to buy green if appealed to appropriately. *Grousers* (19%): Grousers tend to be uneducated about environmental issues and cynical about their ability to effect change. They believe that green products cost too much and do not perform as well as competition. *Basic Browns* (33%): Basic Browns are caught up with day-to-day concerns and do not care about environmental and social issues." Jill Meredith Ginsberg and Paul Bloom, "Choosing the Right Green Marketing Strategy," *MIT Sloan Management Review*, 46/1 (2004): 79–84.

12  The Global Eco-labeling Network (GEN). <http://globalecolabelling.net/> September, 2008.

13  D'Souza Clare *et al.*, "Green Decisions: Demographics and Consumer Understanding of Environmental Labels," *International Journal of Consumer Studies*, 31/4 (2007): 371, and websites of the following Global Eco-Labelling Network op. cit. and ISO. <www.iso.org> May, 2008.

14 OECD Trade Directorate, "CSR and Trade: Informing Consumers about Social and Environmental Conditions of Globalisation Production OECD," *Trade Policy Working Paper* No. 47, Part I.

15 R. Johnston, "A Critique of Life-cycle Analysis: Paper Products," in *The Industrial Green Game: Implications for Environmental Design and Management*, ed. D. J. Richards (Washington, D.C: National Academy of Engineering, 1997): 225–233.

16 R. G. Hunt and R. O. Welch, *Resource and Environmental Profile Analysis of Plastics and Non-plastics Container* (Kansas City: Midwest Research Institute,1974).

17 M. T. Smith, "Squaring the Circle: Fundamental Barriers to Effective Environmental Product Labelling," in *ISO 14001 and Beyond: Environmental Management Systems in the Real World*, ed. C. Sheldon (Sheffield: *Greenleaf* Publishing, 1997): 95–97.

18 This section is based on the report: Katsiaryna Paulavets, *Change and the Food Industry – Climate Labeling for Food Products Potential and Limitations*, TSEL Environmental <http://tsel-environmental.com/> February, 2008, and the notes taken during *the Labeling Climate Change Conference*, Lund, Sweden, November 6–7, 2007, organized by the Global Ecolabelling Network (GEN).

19 Carbon Confusion, *Business Week*, issue 4075, 2008.

20 The Tesco carbon labeling initiative has experienced difficulties in labeling all products in the store. It is uncertain if Tesco will achieve the goal of labeling a 100 percent of its products.

21 Carbon Trust. <www.carbontrust.co.uk> June, 2008.

22 Among the first NGOs working on this was SAFE Alliance now Sustain. <www.sustainweb.org> September, 2008.

23 The WASD has been criticized for not taking into account the mode of transport travelled by products. Another negative aspect of the WASD is the difficulty in using it for products with multiple ingredients.

24 KRAV <www.krav.se> June, 2008.

25 Presentation by Mikael Karlsson during the conference *Labeling Climate Change – Possibilities and Limitations on Labeling Climate Change*, Lund, Sweden, November 6–7, 2007.

26 According to a BBMG Conscious Consumer report, nearly nine in ten Americans say the words *"conscious consumer"* describe them well, and are more likely to buy from companies that manufacture energy efficient products. Raphael Bemporad and Mitch Baranowski, *Conscious Consumers Are Changing the Rules of Marketing. Are You Ready?* (BBMG, November, 2007), <http://www.bbmg.com/pdfs/BBMG_Conscious_Consumer_White_Paper.pdf.>

27 Forest Reinhardt, "Bringing the Environment Down to Earth," *Harvard Business Review*, 77/4 (1999): 149–157.

28 Sadowski and Bukingham Retailers as Choice Editors – European Retail Digest.

29 This section is based on: Renato J. Orsato and Andrea Öström, *"Eco-branding: The Case of Änglamark"*. INSEAD case, 02/2006–5314.

30 This section is based on: Renato J. Orsato, "The Green Building Strategy: The Case of Lend Lease Australia". (UTS, Sydney, Australia, 2006).

31 The building emits only 59 kg of $CO_2$/m2/annum.

32  The Good Consumer, *The Economist*, January 27, 2008.

33  Forest Reinhardt has previously articulated the specific conditions for successful ecological differentiation strategies. For this reason, the examples used in this section only intend
to summarize the considerations that have been sufficiently elaborated by him, and eventually make it easier for those who are not so familiar with the prerequisites of ecologically
differentiation of products. See: Forest Reinhardt, "Environmental Product Differentiation: Implications for Corporate Strategy," *California Management Review*, 40/4 (1998):
43–73. See also: Reinhardt op. cit.; Forest Reinhardt, "Market Failure and the Environmental Policies of Firms: Economic Rationales for 'Beyond Compliance' Behavior," *Journal
of Industrial Ecology*, 3/1 (1999b): 9–21; Forest Reinhardt, *Down to Earth: Applying Business Principles to Environmental Management* (Boston, MA: Harvard Business School Press,
2000): chapter 2.

34  Kontrollföreningen för Ecologisk Odling.

35  For an overview of KRAV, and the general situation of organically grown cereals used for
the production of bread in Sweden, see: Pia Heidenmark, *Going Organic? A Comparative
Study of Environmental Product Development Strategies along Two Swedish Bread Supply
Chains* (Lund, Sweden: IIIEE Dissertations, 2000).

36  James Bessen and Michael J. Meurer, "Do Patents Perform Like Property?," *Academy of
Management Perspectives*, 22/3 (2008): 8–20.

37  Reinhardt, op. cit.

38  The disjunction between intention and actual purchase is discussed by Vogel op. cit.,
Chapter 3; Craig N. Smith, "Consumers as Drivers of Corporate Social Responsibility,"
in The Oxford Handbook of Corporate Social Responsibility, eds. Andrew Crane *et al.*
(New York: Oxford University Press, 2007): 281–302; Ken Peattie and Andrew Crane,
"Green Marketing: Legend, Myth, Farce or Prophesy?," *Qualitative Market Research*, 8/4
(2005): 357–370; Just good business. A special report on corporate social responsibility,
*The Economist*, January, 2008; Esben Rahbek Pedersen and Peter Neergaard, "Caveat
Emptor – Let the Buyer Beware! Environmental Labelling and the Limitations of 'Green'
Consumerism," *Business Strategy and the Environment*, 15/1 (2006): 15–29.

39  The silent tsunami, *The Economist*, June, 2008.

40  Inge Ropke, "The Dynamics of Willingness to Consume," *Ecological Economics*, 28/3
(1999): 399–420.

41  Ropke, op cit.:399.

42  M. Sagoff, "The Allocation and Distribution of Resources," in *The Consumer Society*, eds.
Neva R. Goodwin, Frank Ackeman, and David Kiran (USA: Island Press, 1998): 277–280.

43  S.M. Smith and C.P. Haugtvedt, "Implications of Understanding Basic Attitude Change
Processes and Attitude Structure for Enhancing Pro-environmental Behaviors," in *Envi-
ronmental Marketing: Strategies, Practice, Theory, and Research*, eds. Michael J. Polonsky,
Alma T. Mintu-Wimsatt (New York: The Haworth Press, 1995): 155–176.

## CHAPTER 6 – ENVIRONMENTAL COST LEADERSHIP

1  The Ecolean case has been previously presented in: Renato J. Orsato, "Competitive
Environmental Strategies: When Does It Pay to Be Green? *California Management Review*,
48/2 (2006): 127–141.

2 Information supplied by Mr Per Gustafsson (President Ecolean Group) and Mr Per Gassner (Environmental and Product Director) in March, 2005.

3 The amount of calcium carbonate $3.5 \times 10^{17}$ tones is well distributed around the planet.

4 The life cycle analysis can be downloaded from the Ecolean web site: <www.ecolean.se>.

5 For a discussion of the concept and potential of Eco-design, see Chris Ryan, "Learning from a Decade (or So) of Eco-Design Experience, Part I," *Journal of Industrial Ecology*, 7/2 (2003): 1–12; and Chris Ryan, "Learning from a Decade (or So) of Eco-Design Experience, Part II Advancing the Practice of Product Eco-Design," *Journal of Industrial Ecology*, 8/4 (2005): 3–5.

6 Practical Eco-design methods, tools and guidelines are presented by: Ursula Tischner *et al.*, *How to Do EcoDesign?: A Guide for Environmentally and Economically Sound Design*, 1st edn. (Birkhauser, 2002); John Gertsakis, H. Lewis and C. Ryan, *A Guide to EcoReDesign*. (Melbourne: Centre for Design at RMIT, RMIT University, 1996); H. Lewis and J. Gertsakis. *Design + Environment*. (UK: Greenleaf Books, 2001); J. C Brezet and C. G. van Hemel. *EcoDesign: A Promising Approach to Sustainable Production and Consumption* (Paris: UNEP, 1997).

7 This section is based on: Renato J. Orsato, Frank den Hond, and Stewart R. Clegg, "The Political Ecology of Automobile Recycling in Europe," *Organization Studies*, 23/4 (2002): 639–665; see also: Chapter 11 of Renato J. Orsato, "The Ecological Modernization of Industry: Developing Multi-disciplinary Research on Organization & Environment" (PhD diss., University of Technology, Sydney, 2001).

8 For an overview of the Extended Producer Responsibility (EPR) concept, see: T. Lindhqvist, "Extended Producer Responsibility in Cleaner Production: Policy Principle to Promote Improvements of Product Systems" (PhD diss., University of Lund, 2000).

9 *Bundesminister für Umwelt, Naturschütz und Reactorsicherkeit.*

10 *Verein der Automobilindustrie.*

11 *Projekt Altfahrzeugverwertung der deutschen Automobilindustrie.*

12 The 12 million ELVs generate approximately 2.2 million tons of permanent waste per year. D. Kurylko, "Europe to Regulate Recycling," *Automotive News Europe*, 13 October, 1997, 1; C Wright *et al.*, *"Automotive Recycling: Opportunity or Cost?"* (London: Financial Times Business Limited, 1998).

13 Proposal for a Council Directive on End-of-life Vehicles' (COM 97–358).

14 Directive 2000/53/EC of the European Parliament and of the Council of the 18 September, 2000 on end-of-life vehicles – Official Journal of the European Communities (L269/34).

15 Such assertion is based on discussions within the IPR (Individual Producer Responsibility) working group on 25–26 September, 2008 at INSEAD, Fontainebleau, France. The group is led by Professor Luk Van Wassenhove and formed by academics and company representatives of companies affected by the WEEE Directive.

16 This section was based on Clovis Zapata's "Environmental Regulation and Firm Strategy: The Evolution of Automotive Environmental Policy" (PhD diss., Cardiff University, 2008); Clovis Zapata and Peter Nieuwenhuis, "Driving on Liquid Sunshine – The Brazilian Biofuel Experience: A Policy Driven Analysis, Business Strategy and the Environment" (forthcoming). Peter Wells also contributed to the section.

17  Benjamin Warr and Renato J. Orsato, "Greening the Economy – New Energy for Business Creating a Climate for Change" (paper presented at the European Business Summit 21–22 February, 2008).

18  "Better Living Through Chemurgy," *The Economist*, 26 June, 2008.

19  Jose Goldemberg. P. Coellho, P. Nastari and O. Lucon. "Ethanol Learning Curve – The Brazilian Experience." *Biomass and Bioenergy*, 26 (2004): 301–304.

20  Associacao Nacional dos Fabricantes de Automoveis – ANFAVEA (2008) *Anuario Estatistico da Industria Automobilistica Brasileira*. Sao Paulo: Anfavea.

21  According to P. Nastari *et al.*, *Observations on the Draft Document Entitled Potential for Biofuels for Transport in Developing Countries* (Washington, D.C.: The World Bank, Air Quality Thematic Group. July, 2005) between 1976 and 2004, Ethanol for automotive fuel purposes provided savings of up to US$60,74 billion, in constant dollars of 2005, to the Brazilian government. If external debt interest rate is included in the calculations the figures add up to US$121.26 billion; see also: A. Ashford, "Tanked up on Sugar: Brazil Launched an Experiment to Substitute Alcohol for Oil. How Has It Worked Out?" *New Internationalist* (1989): 95. provides different figures that add up to a less optimistic, though still positive outcome. His analysis calculates savings of US$10.4 billion, between 1975 and 1989, with government subsidies of US$9 billion.

22  J. Magalhaes, N. Kuperman, and R. Machado, "Proalcool: Uma Avaliacao Global" (Sao Paulo Astel Assessores tecnicos, 1992).

23  Vincente Assis, Heinz-Peter Elstrodt and Claudio F.C Silva, "Positioning Brazil for Biofuels Success: The Country Now Produces Ethanol More Cheaply than Anywhere Else on Earth, but that might Not Be True for Long." *McKinsey Quarterly*, Special Edition: Shaping a new agenda for Latin America (2008): 1–5.

24  Luciara Nardon and Kathryn Aten, "Beyond a Better Mousetrap: A Cultural Analysis of the Adoption of Ethanol in Brazil," *Journal of World Business*, 43/3 (July, 2008): 261–273. argue that the success of the Brazilian ethanol program was strongly influenced by socio-cultural elements, which are difficult to be replicated in other contexts.

25  <http://www.brenco.com.br/en/index_en.php> October, 2008.

26  In 1998, C. J. Campbell & J. H. Laherrère, "The End of Cheap Oil," *Scientific American*, 278/3(1998): 60–65 estimate that 1,000 billion barrels of petroleum remained underground, 850 billion in oil reserves, and 150 billion in undiscovered oilfields. At rates of 2 percent growth in consumption per year, the authors predicted that this amount of oil would be able to "fuel" societies for approximately 40 more years (Hence, till 2038). Contrary to popular concern, however, identifying the exact date when the world runs out of petrol is irrelevant. Instead, what is of consequence is when world production will start to decline, or the "Peak Oil" moment. According to the authors, oil production should peak in 2010 and, unless technological and social factors significantly reduce demand for oil, the peak oil will mean increasingly higher oil prices.

27  Jose, Goldemberg, "The Ethanol Program in Brazil," *Environmental Research Letters*, Institute of Physics Publishing edition (2006): sec. Lett: 1.

28  "Better Living Through Chemurgy," *The Economist*, June 26, 2008.

29  NatureWorks' CEO Marc Verbruggen made this comment when interviewed on the Dow Jones newswire in the article: Anjali Cordeiro, "In the Pipeline: Bioplastics Draw Consumer Sector's Attention," *Dow Jones News Service*, October, 2008.

30  <www.natureworksllc.com/> October, 2008.

31 Anjali Cordeiro, "In the Pipeleine: Bioplastics Draw Consumer Sector's Attention," *Dow Jones News Service*. October, 2008.

32 <www.genencor.com> October, 2008.

33 Megan Lampinen, "US: DuPont Danisco Joint Venture Breaks Ground for New Ethanol Facility," *Automotive World* (October, 2008).

34 The information about *eco-n* was collected during the period of 2005–2008 and is documented in a forthcoming INSEAD Teaching case, written by Renato J. Orsato, Richard Christie, Delyse Springett and Ben Warr, The Greening of Pastoral Agriculture: The Case of *Eco-n*.

35 Although agriculture accounts for only 8 percent of the country's GDP, agricultural commodities have historically been the main exports; dairy produce leads with 17 percent, followed by meat, 15.5 percent, forestry products account for 7.1 percent of the exports (Economist Country Briefings, 2005).

36 The science of *eco-n* can be found in: J. Moir, K. C. Cameron, and H. J. Di, "Effects of the Nitrification Inhibitor Dicyandiamide on Soil Mineral N, Pasture Yield, Nutrient Uptake and Pasture Quality in a Grazed Pasture System." *Soil Use Management*, 23 (2007): 111–120; H. J. Di and K. C. Cameron, "Mitigation of Nitrous Oxide Emissions in Spray-irrigated Grazed Grassland by Treating the Soil with Dicyandiamide, a Nitrification Inhibitor." *Soil Use Management*, 19 (2003): 284–290; H. J., Di and K. C., Cameron. "The Use of a Nitrification Inhibitor, Dicyandiamide (DCD), to Reduce Nitrate Leaching from Cow Urine Patches in a Grazed Dairy Pasture under Irrigation." *Soil Use Management*, 18 (2003): 395–403.

37 A hearing for the patent was eventually held by the IPONZ commissioner in September, 2008, but the final outcome was still pending when this book went to press.

38 Under the Kyoto Protocol, NZ is obliged to maintain its GHG emissions during the first commitment period (2008–2012) at 1990 levels. Emissions have increased from 10 MT $CO_2$e in 1990 to 13 MT $CO_2$e in 2003. (Ministry for the Environment (2005) "New Zealand's Grenhouse Gas Inventory 1990–2003" Wellington, New Zealand: New Zealand Government.)

39 *Eco-n* and n-care have the same active ingredient (dicyandiamide – DCD) but *eco-n* is a very fine powder, which is mixed with water and spayed in the pasture with the specific aim of covering the entire soil area, reducing the nitrification effects from randomly deposited urine patches. This is, in fact, the reason for Ravensdown to have applied for a method patent, rather than product – as it was a new use for an existing product. N-care, on the other hand, is in the form of larger particles (a zeolite chip) which incorporates the DCD and is mixed with urea fertilizer.

40 To 11.3 mg N/liter, which is a World Health Organization (WHO) standard.

41 In October, 2008, the price of 1 ton of $CO_2$ was around 23 Euros.

42 Oksana Mont, Singhal Pranshu, and Fadeeva Zinaida, "Chemical Management Services in Sweden and Europe – Lessons for the Future," *Journal of Industrial Ecology*, 10/1–2 (2006): 279–291.

43 <www.chemicalstrategies.org/> October, 2008.

44 More details of the case can be found at: <www.chemicalstrategies.org/case_studies.htm> October, 2008.

45 Oksana Mont, Singhal, Pranshu, and Fadeeva, Zinaida, "Chemical Management Services in Sweden and Europe – Lessons for the Future," *Journal of Industrial Ecology*, 10/1–2 (2006): 279–291.

46  Several examples of PSS can be found in: O. Mont, "Product-Service Systems: Panacea or Myth? Doctoral Dissertation (PhD diss., Lund University, 2004); Chris Ryan, *Digital Eco-sense: Sustainability and ICT – A New Terrain for Innovation* (Lab 3000: Melbourne, Australia, 2004); See also: The Special issue on PSS of *the Journal of Cleaner Production*, 14 (2006); Robin Roy, "Sustainable Product-Service Systems" *Futures*, 32 (2000): 289–299.

47  Basic information about UNEPs program on PSS can be found in <http://www.unep.fr/scp/> *Product-Service Systems and Sustainability – Opportunities for Sustainable Solutions* (United Nations Environmental Programme Division of Technology Industry and Economics Production and Consumption Branch, June, 2000).

48  Robert U. Ayres, "A New Growth Engine for Sustainable Economy," *American Chemical Society*, 32/15 (1998): 367; A. Robert U. Ayres, "Products as Service Carriers: Should We Kill the Messenger – or Send It Back?" (Fontainebleau: INSEAD-UNU CMER, 1999).

49  For an example of and PSS program/network see: <www.suspronet.org> funded by the EU.

50  These theories are: Property rights theory, transaction costs theory, information economics, and principal-agent theory. Gerd Scholl, "Product-Service Systems Taking a Functional and Symbolic Perspective on Usership," in *Perspectives on Radical Changes to Sustainable Consumption and Production, System innovation for sustainability 1*, eds. Arnold Tukker *et al.* (Sheffield: Greenleaf, 2007).

51  The limited empirical evidence of PSS was the main conclusion and topic of debate of the "Second Expert Meeting on Sustainable Innovation: Sustainable Innovation and Business Models", held by UNEP in Paris, 1 October, 2008.

52  Robert Kaplan, *Measures for Manufacturing Excellence* (Boston, MA: Harvard Business School Press, 1991), demonstrated that accounting techniques used by contemporary manufacturing companies are inadequate for the control of modern practices of operations management. According to Kaplan, there has been a *relevant loss* of information about manufacturing processes as a result of the increasing complexity of systems of production. In essence, accounting cost differs greatly from technical costs mainly because the basic principles of financial accounting have practically not evolved over a century. While the general principles of financial accounting have become institutionalized practices in most industrialised countries, technological innovation significantly change the nature of management practices, as well as the organizational structure and processes. Kaplan's critique generated new proposals for measuring the production and performance of companies such as the "activity-based cost" (ABC) accounting, in which costs are broken out and assigned to the causative activities. The advent of ABC systems showed that traditional accounting systems were based on principles and concepts that do not reflect the real performance of organizations.

53  Oksana Mont and Tareq Emtairah, "Systemic Changes and Sustainable Consumption and Production: Cases from Product-Service Systems," in *Perspectives on Radical Changes to Sustainable Consumption and Production, System innovation for sustainability 1*, eds. Arnold Tukker *et al.* (Sheffield: Greenleaf, 2007).

54  Sidney J. Levy, "Symbols for Sale." *Harvard Business Review*, 37/4 (July, 1959): 117–124.

55  Scholl, op. cit.

56  Such logic is in line with the definition of strategy by Michael Porter, "What Is Strategy?" *Harvard Business Review*, 74/6 (1996): 61–79, 61–62; and Joan Magretta, *What Management Is: How It Works and Why It's Everyone's Business* (London: Profile Books, 2003).

## CHAPTER 7 – SUSTAINABLE VALUE INNOVATION

1   In Chapter 9 of Chan W. Kim, and Renée Mauborgne, *Blue Ocean Strategy: How to Create Uncontested Market Space and Make the Competition Irrelevant* (Boston, MA: Harvard Business School Press, 2005) the issue of the "Sustainability of Blue Ocean Strategy" is addressed. However, sustainability there does not refer to environmental or social issues but rather to the BOS itself via barriers to imitation.

2   Paul Hawken, Amory B. Lovins, and L. Hunter Lovins, *Natural Capitalism: The Next Industrial Revolution* (London: Earthscan, 1999).

3   Due to the treatment of "light trucks" in the US there is some confusion about the yearly production of cars. A Land Rover, for instance, is a car in Europe but a light truck in the US. According to the *2008 Global Market Data Book, Automotive News Europe,* the global production of cars in 2007 was 53,190,191 and the global production of trucks was 21,457,069.

4   Paul Niewenhuis, "Developments in Alternative Powertrains", in *Automotive Materials: The Challenge of Globalisation and Technological Change,* ed. Peter Wells (London: Financial Times Automotive (Chapter 9), 1998).

5   J. Womack, D. Jones and D. Roos, *The Machine that Changed the World* (Sydney: Maxwell Macmillan International, 1990).

6   See for instance: R. Badham and J. Mathews, "The New Production Systems Debate." *Labour and Industry,* 2/2 (1989): 194–246.

7   Badham and Mathews, op.cit: 193–197.

8   This includes the use of motor vehicles by one or a few persons only. Throughout the chapter, the terms *personal motorization and individual mobility* also refer to TIMM.

9   Hawken, Lovins, and Lovins, op. cit.

10  A detailed explanation of the concept of *organizational fields* in the context of automobiles is presented by: Renato J. Orsato, "The Ecological Modernization of Industry: Developing Multi-disciplinary Research on Organization & Environment" (PhD diss., University of Technology, Sydney, Australia: 2001).

11  The definition of what constitute a *volume* car manufacturer evolves with the firms' growth of production capacity. Nonetheless, if a production of one million units per year is considered a minimum quantity for a volume manufacturer, the worldwide industry would be comprised by fourteen corporations (*Automotive News Europe:* 2000 Global Market Data Book). Some analysts expect that industry consolidation will result in six global manufacturers with production capacity (volume) at around 15 million cars per year by the year 2020. See: R. Feast, Easy Pickings, *Automotive* World (2000), April: 32–36.

12  GM, Ford, Nissan, Fiat, Saab, Jaguar, Daewoo, Kia, VW, and Mitsubishi.

13  There are, obviously, significant differences between the activities of *manufacturing* the components of an automobile and their subsequent *assembly* into a single unit (motorcar). However, because corporations in the automobile industry have different levels of (vertical) integration between manufacturing and assembling activities, the terms *car or auto manufacturers, carmakers, automakers* and *car or auto assemblers* are used here interchangeably.

14  In the first quarter of 2008, GM and Ford reported losses of $15.5 billion and 8.7 billion, respectively (The Economist, 9 August 2008). In 2009 most carmakers faced heavy financial losses, with Toyota loosing money for the first time in its history.

15 A *break-even point*, as a generic economic concept, represents the quantity where the contribution to fixed costs equals fixed costs. In the auto industry, the break-even point of each car model is the quantity of vehicles that have to be sold to equalize the fixed costs in the same period (normally one year). In theory, then, the break-even is basically about covering fixed costs and these include product development, tooling, plant, etc. In practice, the concept can get quite complicated due to the cost accounting methods. Since carmakers would expect to exceed break-even over the life of a car model, if the life-cycle of the model is 20 years (e.g. Volvo 240), rather than five, they are more likely to exceed the break even. Annual break-even figures are more complex to obtain because all models and production facilities enter the equation. Plant break-even, for instance, refers to the fixed costs for all car models assembled in that factory. For instance, an engine plant that produces engines for several models can break-even, while an assembly run by the same manufacturer, which makes a single unsuccessful model, does not (Source: personal conversation with Paul Niewenhuis).

16 R. Bremner, "Big, Bigger, Biggest", *Automotive World* (2000): 37–43.

17 R. Golding, "Capital Punishment", *Automotive World* (1999): 21.

18 The Economist. Driving into Traffic; March 9, 2006.

19 The fragmentation of markets is illustrated for the case of the UK. In an almost static market in volume terms, the total number of variants on offer more than doubled; from 1,303 in 1994 to 3,155 in 2005.

20 R. J. Orsato and P. Wells, "U-turn: The Rise and Demise of the Automobile Industry," *Journal of Cleaner Production*, 15/11–12 (2007): 994–1006; Peter Wells and Renato J. Orsato, "Redesigning the Industrial Ecology of the Automobile," *Journal of Industrial Ecology*, 9/3 (2005): 15–30; Paul Nieuwenhuis, Philip Vergragt, and Peter Wells, *The Business of Sustainable Mobility: From Vision to Reality* (Sheffield: Greenleaf, 2006); Paul Nieuwenhuis and Peter Wells, *The Death of Motoring?: Car Making and Automobility in the 21st Century* (Chichester: John Wiley & Sons, 1998); Hawken, Lovins, and Lovins, op. cit.

21 Graedel and Allenby in T. Graedel and B. Allenby *Industrial Ecology and the Automobile* (USA: Prentice- Hall International; 1998) for instance, produced a remarkable account of the environmental impact of the entire automotive system. The authors provide detailed information not only about its impact in all phases of its life-cycle, such as energy consumption during car manufacturing as well as details of in-service phases, infrastructure needs, and recycling techniques, but they also suggest alternatives for improving the current system. See, for instance: the research reports of *The Economist Intelligence Unit* on 'The Automobile Industry and the Environment', such as the series of environmental reports of *FT Automotive*. Substantial information is also available on the *Internet*. The Environmental Defence, an American NGO, for instance, presents a basic life-cycle analysis of the impact of automobiles: See: <www.environmentaldefense.org/> May, 2006.

22 J. Whitelegg, *Transport for a Sustainable Future: the Case for Europe* (London: Belhaven; 1993).

23 K. Rogers. *The Motor Industry and the Environment* (London: Report for the Economist Intelligence Unit; 1993).

24 Paul Nieuwenhuis, "Developments in Alternative Powertrains" in: Peter Wells (1998) op. cit.

25 Lovins (1995), op. cit.; Robert Q. Riley, *Alternative Cars in the 21st Century: A New Personal Transportation Paradigm* (USA: SAE International, 1994).

26   Graedel and Allenby op. cit., 114.

27   The use of aluminium for structural car parts and plastic for panels can significantly reduce the weight of a car and, consequently, reduce the emissions. The Audi A2 was the first volume-production car in the world to have a body made entirely of aluminium. The Audi A8 pioneered the use of aluminium in space-frame in 1994. Despite these experiences, the use of aluminium in car structure can be considered marginal in the industry. Lightweight *carbon fiber* can also be used for the panes of cars, but since the global production is still small, carbon fiber is currently too expensive.

28   "Smart Is Fighting on All Fronts," *Automotive news Europe* (April 12 1999).

29   Although the figures for *Smart* are blurred by their consolidation with other figures presented by the Daimler group, the company is not profitable. According to Paul Nieuwenhuis (personal conversation), the break-even of *Smart* is to be located somewhere around 150–160,000 cars/year. Generally models break even at around 65–80 percent of capacity, but it depends on other factors such as margins (small on *Smart*), commonality (low on *Smart*), and length of production (quite good for *Smart ForTwo*). Smart always stated they did not expect to break-even on the first model, expecting the second generation (current For Two) to start making money.

30   This account does not consider, however, the inputs in the form of fertilizers and pesticides, as well as the energy used to process sugarcane into ethanol and then transport it to various refuelling stations. In the end, although bio-fuels are greener than fossil fuels, they are certainly not carbon neutral.

31   Bio-diesel can also contribute to the portfolio of fuels, but the biodiesel market in Brazil is still in its embryonic stages. A governmental program, called the *Programa Nacional de Produçao e Uso de Biodiesel* – PNPB, was established to foster the production and organize the supply chain of the product in 2004. A 2 percent minimum blend to the traditional petrol derived diesel was established. This percentage will increase to 5 percent in 2013 (*Brazilian Law* Number 11,097 of 11 January, 2005).

32   Jose Goldemberg, Suani Teixeira Coelho, and Patricia Guardabassi, "The Sustainability of Ethanol Production from Sugarcane," *Energy Policy*, 36/6 (2008): 2086.

33   "Grow your Own. A Special Report on the Future of Energy," *The Economist*, June, 2008.

34   Vincente Assis, Heinz-Peter Elstrodt and Claudio F. C Silva, "Positioning Brazil for Biofuels Success: The Country Now Produces Ethanol more Cheaply than Anywhere Else on Earth, but that might Not Be True for Long." *McKinsey Quarterly*, Special Edition: Shaping a new agenda for Latin America (2008): 1–5.

35   In the United States of 1900, internal combustion engines powered only 22 percent of the cars; 38 percent were electric and 40 percent were powered by steam engines. The situation changed rapidly by 1905 gasoline-powered automobiles had defeated their competitors. The number of car registrations grew from 8,000 in 1900 to 902,000 in 1912. John Urry, "The 'System' of Automobility," *Theory, Culture & Society*, 21/4–5 (2004): 25–39.

36   For an analysis of the different powertrain options, see: Paul Nieuwenhuis, Philip Vergragt, and Peter Wells, *The Business of Sustainable Mobility: From Vision to Reality* (Sheffield: Greenleaf, 2006).

37   For a detailed description of *hybrid electric vehicles*, see: Hawken, Lovins, and Lovins, *op. cit.*, Victor Wouk, "Hybrid Electric Vehicles," *Scientific American*, 277/4 (1997): 70–75; Daniel Sperling, *Future Drive: Electric Vehicles and Sustainable Transportation*, 1st ed. (Washington D.C: Island Press, 1994), Chapter 6; Amory Lovins, "Moving toward a

New System", in: *Building the E-Motive Industry: Essays and Conversations about Strategies for Creating an Electric Vehicle Industry*, ed. S. A. Cronk (USA: SAE International, 1995).

38  *Concept cars* are prototypes that point to future technological choices; "They Narrow the Pool of Technologies from which Manufacturers Are Likely to Choose and as Such Are a Valuable Indication of Future Trends"; Paul Nieuwenhuis and Peter Wells, *The Death of Motoring?: Car Making and Automobility in the 21st Century* (Chichester: John Wiley & Sons, 1998).

39  Between 1997–2007, Toyota sold more than one million hybrid vehicles worldwide. Prius represents more than 90 percent of the sales, which includes also the hybrids Highlander and Camry. More than half of these vehicles (557.276 units) were sold in North America. (See: Toyota North America Environmental Report, 2007, 16–17).

40  In 2008, besides the hybrids of Toyota (Prius and Camry) and Lexus GS 450h, Lexus LS 600h, which are also owned by Toyota, among other commercial hybrids were: Honda Civic, Nissan Altima, Ford Escape, Mazda Tribute, and Saturn Aura. Many more were expected to be launched in the coming years.

41  "Tesla's Wild Ride," *Fortune*, 21 July, 2008.

42  Renato J. Orsato, "Future Imperfect? The Enduring Struggle for Electric Vehicles," in Nieuwenhuis, Vergragt, and Wells op. cit.

43  Nieuwenhuis and Wells (1998), op. cit., 97.

44  Around one million Smart ForTwo cars were sold between 1998 and 2007, making an average of 100,000 cars per year. Italy is the largest market, followed by Germany and more recently the US with 16,000 cars sold between January and September, 2008. <www.smart.com> and <www.media.daimler.com> September, 2008.

45  Joan Magretta, *What Management Is: How It Works and Why It's Everyone's Business*, 1st ed. (New York: Free Press, 2002).

46  Jonathan Lash and Fred Wellington, "Competitive Advantage on a Warming Planet," *Harvard Business Review*, 85/3 (2007): 94–102.

47  A detailed explanation of the *all steel-body monocoque paradigm* is provided by: Paul Nieuwenhuis and Peter Wells, "The All-steel Body as a Cornerstone to the Foundations of the Mass Production Car Industry." *Industrial and Corporate Change* 16/2 (2007): 183–211; Paul Nieuwenhuis and Peter Wells, *The Automotive Industry and the Environment: A Technical, Business and Social Future* (Boca Raton: CRC Press, 2003); Paul Nieuwenhuis and Peter Wells, *The Death of Motoring?: Car Making and Automobility in the 21st Century* (Chichester: John Wiley & Sons, 1998).

48  According to Nieuwenhuis and Wells op. cit.; the *Buddist* paradigm involves high investments in the press shop (where steel sheet is pressed into shaped panel), body shop (where the pressed panels are welded together into subassemblies and thence into a monocoque body-chassis unit) and paint shop before any products have been developed. The minimum investment in a press shop would be around €160 million. Body shop investment depends on automation levels, but can be between €80 million and €160 million.

49  Renault's Logan has been a great sales success. Since its launch in 2004 till September, 2008, Renault sold more than one million units, or around 250,000 cars per year. <www.renault.com> September, 2008.

50  The concept used here differs from the one referring to *multi-purpose vehicles* (MPV) used in classifications of market segmentation to denominate a specific market niche.

51  Three market segments, denominated 'core segments', account for approximately 70 percent of total sales of automobiles in Europe. The other 30 percent comprise niche market vehicles, such as four-wheel drive, luxury and sports vehicles.

52  Wayt W. Gibbs, "Transportation's Perennial Problems," *Scientific American*, 277/4 (1997): 54–58.

53  Peter Wells, "The Limits to Growth," *Automotive Environmental Analyst* (2004). Besides congesting, there are serious concerns about the availability of key materials used in cars.

54  According to United Nations "World Population to 2300" (2004) the estimated number of population for 2020 is 7.5 billions (p. 5, medium scenario).

55  Among the authors discussing how business models influences performance are: Christoph Zott and Raphael Amit, "The Fit Between Product Market Strategy and Business Model: Implications for Firm Performance," *Strategic Management Journal*, 29/1 (2008): 1–26; Christoph Zott et Raphael Amit, "Business Model Design and the Performance of Entrepreneurial Firms," *Organization Science*, 18/2 ( 2007): 181–199; Joan Magretta, "Why Business Models Matter," *Harvard Business Review*, 80/5 (2002): 86–92; George S. Yip, "Using Strategy to Change Your Business Model," *Business Strategy Review*, 15/2 (2004): 17–24; Henry Chesbrough and Richard S. Rosenbloom, "The Role of the Business Model in Capturing Value from Innovation: Evidence From Xerox Corporation's Technology Spin-off Companies," *Industrial & Corporate Change*, 11/3 (2002): 529–555; Michael Morris, Minet Schindehutte, and Jeffrey Allen, "The Entrepreneur's Business Model: Toward a Unified perspective," *Journal of Business Research*, 58/6 (2005): 726–735; Stephen L. Vargo and Robert F. Lusch, "Evolving to a New Dominant Logic for Marketing," *Journal of Marketing*, 68/1 (2004): 1–17.

56  Morgan: <www.morgan-motor.co.uk/>, Lotus: <www.grouplotus.com>, and Axon Automotive: <www.axonautomotive.com> September, 2008.

57  <www.agglo-larochelle.fr> August, 2008.

58  BART was later converted into *CarLink* system. For a case study on CarLink see: <http://cat.inist.fr/> October, 2008.

59  'Individual' or 'personalized' transport does not mean a single individual using a vehicle, but rather includes as many persons as a car can carry, which in average is around five individuals.

60  See, for instance: Jochen Markard and Bernhard Truffer, "Technological Innovation Systems and the Multi-level Perspective: Towards an Integrated Framework," *Research Policy*, 37/4 (2008): 596; See also the classic article of Giovanni Dosi, "Technological Paradigms and Technological Trajectories – A Suggested Interpretation of the Determinants and Directions of Technical Change," *Research Policy*, 22/2 (1993): 102–103.

61  Clayton M. Christensen, *The Innovator's Dilemma: The Revolutionary Book that Will Change the Way You Do Business* (New York: Harper Business Essentials, 2003), page xvi.

62  Andrew Williams, "Product Service Systems in the Automobile Industry: Contribution to System Innovation?" *Journal of Cleaner Production*, 15/11–12 (2007): 1093–1103.

63  The section about Better Place was developed with material from interviews with the Better Place leadership team from August to October, 2008 – Shai Agassi (President and CEO), Tal Agassi (Global Deployment), Alizia Peleg (Head of Global Operations), Barak

Hershkovitz (Operating Systems), Joe Paluska (Head of Marketing), Sidney Goodman (Strategic Alliances and Partnerships), Josh Steinmann (Head of Business Development North America and Europe), and Moshe Kaplinsky (CEO Better Place Israel). An expanded version of this case can be found in Renato J. Orsato and Sophie Hemne "Better Place: Sustainable Value Innovation in Mobility". INSEAD Teaching Case (forthcoming).

64  Deutsche Bank Estimates that Lithium Ion batteries, depending on which type, will cost around US$500–600/kWh which comes to US$11,000 for a full EV 22kWh. Deutsche Bank, *Electric Cars: Plugged In-Batteries must Be Included*, 9 June, 2008.

65  A Deutsche Bank report predicts that the Lithium Ion battery market will boom by 2020. However, there are concerns about safety and performance, including thermal runway, over-charging, difficulties to operate in extremes weather conditions, as well as battery durability and deterioration. Deutsche Bank, *Electric Cars: Plugged In-Batteries must Be Included*, 9 June, 2008.

66  This argument is based on a market survey developed by Better Place and *Transportvaneundersøgelsen*. The results are valid for Israel and Denmark.

67  Israel imposed a 72 percent tax on ICE and Denmark 180 percent.

68  D. Yergin *The Prize* (New York: Touchstone Books, 1992).

69  Williams, op. cit.; Franz E. Prettenthaler and Karl W. Steininger, "From Ownership to Service Use Lifestyle: The Potential of Car Sharing," *Ecological Economics*, 28/3 (1999): 443–453; Susan Shaheen, "Commuter Based Car-sharing – A market Niche Potential" *Transportation Research Record,* 1760 Paper No. 01:3055.

70  Information about the history of *Mobility Switzerland* can be found in Hoogma, Kemp, Schot and Truffer, *Experimenting for Sustainable Transport.*

71  Auto Teilet Genossenschaft, ATG, and ShareCom.

72  <www.mobility.ch> July, 2008.

73  <www.zipcar.com> September, 2008.

74  See Endnote 26 in Chapter 6.

75  The trend towards CSR is well documented in a vast literature. For an overview see: Andrew Crane, Abagail McWilliams, Dirk Matten, Jeremy Moon, and Donald S. Siegel, *The Oxford Handbook of Corporate Social Responsibility* (New York: Oxford University Press, 2008); See also: Wayne Visser *et al.*, *The A to Z of Corporate Social Responsibility: A Complete Reference Guide to Concepts, Codes and Organisations* (Wiley, 2008).

76  Path 6, look across time, Kim and Mauborgne op. cit., 75–79.

77  Susan Shaheen and Adam Cohen, "Worldwide Carsharing Growth: An International Comparison," *Transportation Research Record*, No. 1992, (2007): 81–89. See also: <www.carsharing.net/library/index.html> September, 2008.

78  Prettenthaler and Steininger, op. cit.

79  Williams, op. cit., Prettenhaler and Steininger, op. cit.

80  This argument is based on a market survey developed by Better Place and *Transportvaneundersøgelsen*. The results are valid for Israel and Denmark.

81  Gerd Scholl, "Product-service systems taking a functional and symbolic perspective on usership" in Arnold Tukker *et al. Perspectives on Radical Changes to*

*Sustainable Consumption and Production, System Innovation for Sustainability 1* (Sheffield: Greenleaf, 2007)

82 Prettenthaler and Steininger, op. cit.

83 A 2006 survey in China revealed that 25 percent of the respondents expressed a high interest in car-sharing services. Susan A. Shaheen and Eliot Marting, "Demand for Car-sharing Services in China: An Assessment of Shared-Use Vehicle Market Potential in Beijing," *International Journal of Sustainable Transportation,* forthcoming.

84 See Chapter 5 in Kim and Mauborgne, op. cit.

85 According to the survey developed for *Velib,* more than half of the users in spring 2007 had been using the bikes since their inauguration in July, 2007. According to the same customer survey, 94 percent of the users were satisfied with the service put in place and 97 percent would recommend it to a friend. Source: <http://www.velib.paris.fr/> September, 2008.

86 The City of Paris is currently in the process of developing a similar system for 2000 Electric cars called Autolib'. The scheme will be developed using the same logic of the Velib system (FR3 7 October, 2008).

87 According to Susan Shaheen (personal communication, October, 2008), since the 1990s, some carmakers have been active in pilot projects about car-sharing. In Asia, Honda has been involved in ICVS and Diraac (now KahShare and operated by Kah motors), and Toyota in Toyota-Crayon. Honda has also been involved in CarLink I and II and Intellishare and ZEV-NET in the US. Other auto companies have researched car-sharing including Mercedes Benz, Ford, and Nissan.

88 At *Mobility Switzerland,* 80,000 members share 3,000 cars (one car for 26.6 persons), and 200,000 clients share 5,000 cars in *Zipcar* (one car for 40 persons).

89 Clayton M. Christensen, Scott Cook, and Taddy Hall, "Marketing Malpractice," *Harvard Business Review,* 83/12 (2005): 74–83.

90 The term *flexible specialization* was used to describe the Italian economic miracle with industrial clusters of small and medium enterprises in the 1980s. See: Michael J. Piore, *The Second Industrial Divide: Possibilities for Prosperity* (New York: Basic Books, 1984).

91 Among them are: *Curves* (a Texan-based women fitness company); *Netjets* (a time-sharing jet company); *Cirque du Soleil* (entertainment); *Southwestern* airlines, *Bloomberg* (financial software), NABI (municipal busses), and *Corporate Foreign Exchange* (online database systems).


## CHAPTER 8 – SUSTAINABILITY STRATEGIES AND BEYOND

1 Paul R. Lawrence and Nitin Nohria, *Driven: How Human Nature Shapes Our Choices* (Jossey-Bass, 2002).

2 Reinhardt, op. cit.

3 A useful review of Insert Management Innovation is presented by: Julian Birkinshaw, Gary Hamel, and Michael J. Mol, "Management Innovation," *Academy of Management Review,* 33/4 (2008): 825–845.

4 Report of the World Commission on Environment and Development: "Our Common Future," 1987. (Also known as the Brundtland report.)

5   Lester R. Brown, *Plan B 2.0: Rescuing a Planet Under Stress and a Civilization in Trouble*, Exp Upd. (W. W. Norton, 2006).

6   Paul Hawken, Amory Lovins, and L. Hunter Lovins, *Natural Capitalism: Creating the Next Industrial Revolution* (Back Bay Books, 2000).

7   An excellent review of Deep ecology has been made by Fritjof Capra, *Web of Life* (UK: Harper Collins, 1997).

8   Renato J. Orsato and Stewart R. Clegg, "Radical Reformism: Towards Critical Ecological Modernization," *Sustainable Development*, 13/4 (2005): 253–267.

# INDEX